AF410862

Ecology of Predation and Scavenging and the Interface

Ecology of Predation and Scavenging and the Interface

Editor

Marcos Moleón

MDPI • Basel • Beijing • Wuhan • Barcelona • Belgrade • Manchester • Tokyo • Cluj • Tianjin

Editor
Marcos Moleón
University of Granada
Spain

Editorial Office
MDPI
St. Alban-Anlage 66
4052 Basel, Switzerland

This is a reprint of articles from the Special Issue published online in the open access journal *Diversity* (ISSN 1424-2818) (available at: https://www.mdpi.com/journal/diversity/special_issues/ Predation_Scavenging).

For citation purposes, cite each article independently as indicated on the article page online and as indicated below:

LastName, A.A.; LastName, B.B.; LastName, C.C. Article Title. *Journal Name* **Year**, *Volume Number*, Page Range.

ISBN 978-3-0365-1040-8 (Hbk)
ISBN 978-3-0365-1041-5 (PDF)

Cover image courtesy of Sergio Eguía

Contents

About the Editor

Marcos Moleón is member of the research and teaching staff of the Department of Zoology at the University of Granada (Spain), close to his birthplace. He is a biologist and obtained his PhD on predator–prey interactions in 2007. Since then, his research has mostly focused on the ecology of predation and scavenging, as well as their interface. A major transversal aim of his research is to provide science-based arguments to solve outstanding conservation problems of vulnerable wildlife populations and systems. His main study models are birds of prey and mammalian carnivores, especially in the Mediterranean Basin and the African continent. He is also passionate about human evolution and the present and past relation between our species and megafauna. For more information, you may visit his personal website at http://wpd.ugr.es/mmoleon/.

Acknowledgments

I would like to sincerely thank all the authors for their remarkable contribution to this book, the peer reviewers for their time and constructive spirit, and José A. Sánchez-Zapata for critically reviewing the Preface and Introduction. Also, I wish to thank the staff members at the MDPI editorial office for their dedicated support. Finally, I'm grateful to the Spanish Ministries concerned with science and EU ERDF funds for funding my research contract RYC-2015-19231 and project CGL2017-89905-R.

Dedication: To my mother, to her trusting and serene gaze.

Preface to "Ecology of Predation and Scavenging and the Interface"

Imagine the rise of animals on our planet. Around 650 million years ago, certain multicellular organisms began to thrive in the waters and on seabed of the primitive ocean. These earliest animals primarily fed on non-animal organisms, either living or dead. However, these animals themselves represented an additional food resource, one that was particularly nutritional. Straight away, we can also envisage how some of those tiny animals promptly started to exploit this emergent resource, thus leading to the first carnivory events ever on Earth. In those rudimentary aquatic ecosystems, the earliest carnivorous species very likely exploited animals—possibly transported by currents or resting on the seabed—that could not escape from being consumed, because of their lack of mobility or simply because they were dead. During the Cambrian explosion, around 100 million years later, some of these creatures, perhaps in an attempt to avoid predators, forged their way onto land, under a low-oxygen atmosphere. Amongst the ancient arthropods that ventured into this newly occupied and challenging environment, there were carnivorous species that undoubtedly made use of other live and dead animals. Thus, carnivory surely displayed its predation–scavenging duality from the very first moment of animal history. Since then, many species in most faunal groups have adopted a carnivorous diet, either strict or opportunistic.

When the first naturalists and scientists, in their initial efforts to unravel nature, engaged in the arduous task of disentangling and naming its basic components, they usually considered predators (and predation) to be neatly different from scavengers (and scavenging). This distinction has pervaded the ecological and evolutionary research of carnivorous animals until very recently, which, admittedly, still causes me some surprise. At the beginning of the 21st century, the pioneer work of Travis L. DeVault and his colleagues, in North America, and Nuria Selva and her colleagues, in Europe, called broad attention to the fact that both processes, predation and scavenging, were indeed intimately linked. These and other studies widely showed that there are not pure predators and pure scavengers, but that different individuals, populations and species are positioned somewhere in the predation–scavenging gradient. For a carnivorous animal, including hominins, the decision of hunting prey or scavenging its carcass is merely the result of the balance between the costs in accessing the food and its energetic reward, plus some aspect of chance—exactly like that which characterizes many other biological phenomena.

By that time, I was facing a new, largely uncertain, though decisive, stage in my life. Like many other young researchers, once I finished by PhD thesis on predator–prey interactions, I was eagerly struggling to find my place in the postdoctoral universe. The mentioned pioneer works became a real stimulus in guiding my research scheme onwards. The stimulating, enthusiastic, informal, and always funny conversations with my good friend José A. Sánchez-Zapata, along with inspiration from other researchers that I also profoundly admired, such as José A. Donázar, Fernando Hiraldo, and Norman Owen-Smith, greatly motivated me to begin the basis of a project to explore the interactions between predators and scavengers. Not surprisingly, the first steps of this novel research line for me were done in Africa. This continent meets all the ideal conditions for this purpose, as it still maintains (unfortunately, in an increasingly restricted number of areas) complex assemblages of carnivorous species, including large predators and vultures, as well as a varied offer of prey and carrion, including megaherbivores. In Africa, moreover, one can feel himself, probably like in no other place, as an integral part of nature, which aids the process of thinking about the ecological interactions that take place within food webs—and that shaped our origin and evolution as humans.

Today, we know well that full recognition of the dual facet of carnivorous species has profound implications at different levels of life organization. However, most of them still remain largely unexplored. Some of these relevant questions include: What is the proportion of predated versus scavenged prey in the diet of carnivorous animals? Which are the factors that make some individuals more prone to scavenging, and some others more prone to hunting? To what extent do predators and scavengers compete for shared resources? Can predators benefit scavengers, and vice versa? How may these interactions between predators and scavengers indirectly affect prey populations? What is the role of carcasses as information centers and predator and parasite risk sites for scavengers, predators and their prey? How may the interaction between predation and scavenging challenge established principles in food web and community structure and function? Are there new methodological tools to help advance our understanding of the interface between predation and scavenging?

This book is the result of a Diversity's Special Issue, of which I was honored to guest edit, which was an exciting opportunity to deal with some of these and other questions. The book's main goal is to consolidate our awareness of the close connection that exists between predation and scavenging. Through its six chapters, I hope this compendium of science may help to inspire ecologists, evolutionary biologists, paleontologists, anthropologists, epidemiologists, forensic scientists, anatomists, and, of course, conservation biologists in their stimulating and promising endeavor of achieving a more comprehensive understanding of carnivory in this rapidly changing world.

Marcos Moleón
Editor

Editorial

Ecology of Predation and Scavenging and the Interface: A Special Issue

Marcos Moleón

Department of Zoology, University of Granada, 18071 Granada, Spain; mmoleonpaiz@hotmail.com or mmoleon@ugr.es

Citation: Moleón, M. Ecology of Predation and Scavenging and the Interface: A Special Issue. *Diversity* **2021**, *13*, 95. https://doi.org/10.3390/d13020095

Received: 18 February 2021
Accepted: 19 February 2021
Published: 22 February 2021

Publisher's Note: MDPI stays neutral with regard to jurisdictional claims in published maps and institutional affiliations.

Predation and scavenging are pervasive ecological interactions in both terrestrial and aquatic environments. The ecology, evolution and conservation of scavengers, and especially predators, have received wide scientific attention and public awareness. However, the close connection that exists between predation and scavenging has not been made explicit until recently [1–3]. The propensity to hunt or scavenge a prey may vary within individuals, among different individuals within a population, and among different populations and species, depending on an intricate array of both intrinsic (e.g., morphology, body condition) and extrinsic (e.g., availability of alternative food sources) factors. In turn, the recognition that carnivorous animals may obtain meat by either hunting prey or scavenging their carcasses has profound implications, from individual morphology, physiology, and behavior to population, community, and ecosystem structure and functioning [1–5].

Given the novelty of this integrative research topic, many relevant questions have yet to be resolved. This Special Issue, through the three research papers and the three reviews that comprise it, aims to deal with some of these questions from diverse perspectives and methodological approaches.

In the first paper of this SI, Ordiz et al. [6] describe, in detail, the predatory and scavenging behavior of wolves (*Canis lupus*) and bears (*Ursus arctos*) in a Swedish area to understand the intrinsic and extrinsic conditions that favor the coexistence of these competing top carnivores. They show that bears and wolves were connected by frequent indirect interactions, mainly through bear scavenging of wolf kills. Scavenging by bears diminished in the moose calving season, when both carnivores turned to the abundant and vulnerable calves as the main food source. Additionally, not all bears were equally prone to scavenging wolf kills, as these carcasses were avoided by females with cubs of the year, i.e., the bear population sector that is more vulnerable to predation.

Teurlings et al. [7] explore a major anti-scavenger strategy of the other top carnivore, the Eurasian lynx (*Lynx lynx*). To prevent scavenger access to the remains of large prey, and thus to secure subsequent meals, lynxes and other felids usually hide their kills by covering them with different materials, such as vegetation and snow. This study, conducted in an area of southeastern Norway, shows that this caching behavior is an efficient anti-scavenger strategy, as cached prey (namely, roe deer, *Capreolus capreolus*) were discovered later than non-cached prey by both vertebrate (especially, birds) and invertebrate scavengers. These results are crucial to fully explain the functional responses of lynxes to their prey, and lynx–prey dynamics in general.

In the next empirical study, Teurlings et al. [8] further focus on the Eurasian lynx–roe deer system to investigate whether above-ground ecological processes linked to predation can trigger cascading effects on below-ground processes via carrion supply and decomposition. Unlike similar studies conducted in other systems, Teurlings et al. did not detect any effect of carcass remains on key chemical parameters of soil and vegetation about two years after death. These findings could be explained by the relatively small size of roe dear carcasses and by their efficient consumption by lynxes and scavengers.

The first review of this SI, made by Luna et al. [9], compares the scientific effort that has been devoted to date to predation and scavenging processes in urban habitats, which are increasingly represented in the planet Earth. The authors found that predation has been far more studied than scavenging. Moreover, urban ecologists became interested in scavenging several decades later than predation. The study species and areas of articles on scavenging were a subset of the species and areas studied in articles on predation. Luna et al. conclude that proper recognition of both the predatory and scavenging facets of carnivores will be needed to fully understand their role in urban food webs and their ecological consequences for urban environments.

A key question in predation–scavenging research is to identify the adaptations that make a species successful in exploiting a given niche within the predation–scavenging gradient. In this line, Potier [10] reviews the visual specializations associated with predatory and scavenging diurnal raptors. He finds that the eye size relative to body mass, as well as binocularity (as opposed to an enlarged field of view), increases towards the predation extreme of the gradient. He also identifies a qualitative anatomical difference between typical predators and more opportunistic and scavenger species, with the former having a second, temporally positioned fovea (probably used during prey capture) in addition to the central fovea that occurs in all species. These findings highlight the close relationship between visual system specializations and foraging ecology, which was often unrelated to phylogeny.

In the last contribution to this SI, Moleón and Sánchez-Zapata [11] reveal the important, though largely overlooked, role that carrion plays in the landscapes of fear and disgust. By reviewing the scientific literature, they identify the main ways in which carrion may be scary and disgusting, namely the principal interaction pathways between carcasses and their visitors (both carnivore and herbivore species) that expose the former to predators (see Ordiz et al. [6] for an empirical example in this SI) and parasites at carcass sites. In addition, they identify major knowledge gaps, which are mostly related to the disgusting facet of carrion. The presented conceptual framework may help to understand animal behavior and ecological processes, including cascading effects, around carrion resources.

The papers and reviews of this SI are proof of the explicit interest in the relationship between predation and scavenging that has currently pervaded many research groups worldwide. Nevertheless, as evidenced by this SI, important knowledge gaps still arise. For instance, investigations into marine, freshwater, and tropical terrestrial environments, as well as on invertebrates, would be especially welcome. I hope this SI may contribute to inspire future research ideas and effort on this general topic. Overall, the growing body of scientific knowledge on the interface between predation and scavenging will definitely dismiss the traditional view that they are disconnected ecological processes.

Funding: The author was supported by a research contract Ramón y Cajal from the Spanish Ministry of Economy and Competitiveness (RYC-2015-19231) and by a project from the Spanish Ministry of Economy, Industry and Competitiveness and EU ERDF funds (CGL2017-89905-R).

Acknowledgments: I would like to thank all the authors for their remarkable contribution to this SI. I also wish to thank the staff members at the MDPI editorial office for their dedicated support.

Conflicts of Interest: The author declares no conflict of interest.

References

1. De Vault, T.L.; Rhodes, O.E., Jr.; Shivik, J.A. Scavenging by vertebrates: Behavioral, ecological, and evolutionary perspectives on an important energy transfer pathway in terrestrial ecosystems. *Oikos* **2003**, *102*, 225–234. [CrossRef]
2. Wilson, E.E.; Wolkovich, E.M. Scavenging: How carnivores and carrion structure communities. *Trends Ecol. Evol.* **2011**, *26*, 129–135. [CrossRef] [PubMed]
3. Moleón, M.; Sánchez-Zapata, J.A.; Selva, N.; Donázar, J.A.; Owen-Smith, N. Inter-specific interactions linking predation and scavenging in terrestrial vertebrate assemblages. *Biol. Rev.* **2014**, *89*, 1042–1054. [CrossRef] [PubMed]
4. Barton, P.S.; Cunningham, S.A.; Lindenmayer, D.B.; Manning, A.D. The role of carrion in maintaining biodiversity and ecological processes in terrestrial ecosystems. *Oecologia* **2013**, *171*, 761–772. [CrossRef] [PubMed]
5. Pereira, L.M.; Owen-Smith, N.; Moleón, M. Facultative predation and scavenging by mammalian carnivores: Seasonal, regional and intra-guild comparisons. *Mammal Rev.* **2014**, *44*, 44–45. [CrossRef]

6. Ordiz, A.; Milleret, C.; Uzal, A.; Zimmermann, B.; Wabakken, P.; Wikenros, C.; Sand, H.; Swenson, J.E.; Kindberg, J. Individual variation in predatory behavior, scavenging and seasonal prey availability as potential drivers of coexistence between wolves and bears. *Diversity* **2020**, *12*, 356. [CrossRef]
7. Teurlings, I.J.M.; Odden, J.; Linnell, J.D.C.; Melis, C. Caching behavior of large prey by Eurasian lynx: Quantifying the anti-scavenging benefits. *Diversity* **2020**, *12*, 350. [CrossRef]
8. Teurlings, I.J.M.; Melis, C.; Skarpe, C.; Linnell, J.D.C. Lack of cascading effects of Eurasian lynx predation on roe deer to soil and plant nutrients. *Diversity* **2020**, *12*, 352. [CrossRef]
9. Luna, Á.; Romero-Vidal, P.; Arrondo, E. Predation and scavenging in the city: A review of spatio-temporal trends in research. *Diversity* **2021**, *13*, 46. [CrossRef]
10. Potier, S. Visual adaptations in predatory and scavenging diurnal raptors. *Diversity* **2020**, *12*, 400. [CrossRef]
11. Moleón, M.; Sánchez-Zapata, J.A. The role of carrion in the landscapes of fear and disgust: A review and prospects. *Diversity* **2021**, *13*, 28. [CrossRef]

Article

Individual Variation in Predatory Behavior, Scavenging and Seasonal Prey Availability as Potential Drivers of Coexistence between Wolves and Bears

Andrés Ordiz [1,2,3,*], Cyril Milleret [1], Antonio Uzal [3], Barbara Zimmermann [4], Petter Wabakken [4], Camilla Wikenros [2], Håkan Sand [2], Jon E Swenson [1] and Jonas Kindberg [5,6]

[1] Faculty of Environmental Sciences and Natural Resource Management, Norwegian University of Life Sciences, Postbox 5003, NO-1432 Ås, Norway; cyril.milleret@gmail.com (C.M.); jon.swenson@nmbu.no (J.E.S.)

[2] Grimsö Wildlife Research Station, Department of Ecology, Swedish University of Agricultural Sciences, SE-730 91 Riddarhyttan, Sweden; camilla.wikenros@slu.se (C.W.); hakan.sand@slu.se (H.S.)

[3] School of Animal, Rural and Environmental Sciences, Nottingham Trent University, Brackenhurst, Southwell, Nottinghamshire NG25 0FQ, UK; antonio.uzal@ntu.ac.uk

[4] Faculty of Applied Ecology, Agricultural Sciences and Biotechnology, Inland Norway University of Applied Sciences, Evenstad, NO-2480 Koppang, Norway; barbara.zimmermann@inn.no (B.Z.); petter.wabakken@inn.no (P.W.)

[5] Norwegian Institute for Nature Research, NO-7485 Trondheim, Norway; jonas.kindberg@rovdata.no

[6] Department of Wildlife, Fish, and Environmental Studies, Swedish University of Agricultural Sciences, SE-901 83 Umeå, Sweden

* Correspondence: andres.ordiz@gmail.com

Received: 11 August 2020; Accepted: 12 September 2020; Published: 15 September 2020

Abstract: Several large carnivore populations are recovering former ranges, and it is important to understand interspecific interactions between overlapping species. In Scandinavia, recent research has reported that brown bear presence influences gray wolf habitat selection and kill rates. Here, we characterized the temporal use of a common prey resource by sympatric wolves and bears and described individual and seasonal variation in their direct and/or indirect interactions. Most bear–wolf interactions were indirect, via bear scavenging of wolf kills. Bears used >50% of wolf kills, whereas we did not record any wolf visit at bear kills. Adult and subadult bears visited wolf kills, but female bears with cubs of the year, the most vulnerable age class to conspecifics and other predators, did not. Wolf and bear kill rates peaked in early summer, when both targeted neonate moose calves, which coincided with a reduction in bear scavenging rate. Some bears were highly predatory and some did not kill any calf. Individual and age-class variation (in bear predation and scavenging patterns) and seasonality (in bear scavenging patterns and main prey availability of both wolves and bears) could mediate coexistence of these apex predators. Similar processes likely occur in other ecosystems with varying carnivore assemblages.

Keywords: apex predators; bear; interspecific interactions; moose; predation; scavenging; wolf

---

1. Introduction

During the last two centuries, large carnivores have suffered drastic population declines, range contractions, and habitat fragmentation [1]. Although carnivores have adapted to almost every habitat, barely ~5% of the Earth's terrestrial land area contains five or more overlapping species of large carnivores [2]. Despite rarity and typically low population densities, large carnivores influence ecosystems in multiple ways through predator-prey interactions, i.e., carnivores are keystone species [3].

They affect prey and mesopredators in both demographic and behavioral terms, which can ultimately drive trophic cascades [4].

Although many large carnivore populations remain threatened [2], others are expanding nowadays in different continents [5,6], increasing the chances for different species to overlap. At the worldwide scale, northern Eurasia is the region with the greatest expansion range of a four-species guild (gray wolves *Canis lupus*, Eurasian lynx *Lynx lynx*, brown bear *Ursus arctos,* and wolverines *Gulo gulo*) [2,6]. Natural recolonization by large carnivores provides opportunities to study interspecific interactions among them, which is crucial to understanding how they can affect each other and lower trophic levels.

Several empirical studies on interspecific interactions among large carnivores have been conducted in northern Europe. Research has focused on habitat and resource use by different species in the large carnivore guild [7]; competition between species, such as wolf and lynx [8] and wolverine and lynx [9]; interference competition between trophic levels [10]; and on the demographic impact of coexisting predators on prey [11]. Interspecific interactions between the two largest carnivores (brown bears and wolves) that roam over large areas of the Northern Hemisphere have been studied in North-America (e.g., [12–14]) and in Scandinavia (Norway and Sweden) in recent years [15]. In Scandinavia, research has focused on wolf habitat selection at different spatial scales [16–19] and on the wolves' kill rates in areas sympatric and allopatric with bears [20]. In moose *Alces alces*—bear—wolf systems, predation is a major driver of moose population dynamics [21], so research has also informed management to optimize ungulate harvest yield where wolves and bears coexist [22].

The body of literature on large carnivores in northern Europe has highlighted the existence of individual variation in habitat selection and kill rates. For instance, wolverines display high individual variation when selecting home ranges [23] and lynx show individual variation in home-range size [8]. Large individual variation has also been shown in wolf home range size [24], bear habitat selection [25], and bear kill rates [26]. The latter implies different levels of specialization reflecting individual foraging behavior [27]; in turn, individual differences in predator behavior may help explain the large individual variation in bear habitat selection [17].

Intraspecific variation in habitat selection may be an adaptation for wolves and bears to reduce intra- and interspecific competition, i.e., individual variation may promote coexistence between these large predators [17]. The role of individual variation in habitat selection, activity patterns [28], and foraging behavior [27] at higher levels of biological organization is indeed gaining increasing recognition [29,30]. Individual variation can have consequences for population and community ecology [31], favoring coexistence of sympatric species [32].

Besides individual variation, seasonality is another important factor that may have implications for coexistence between large carnivores [33]. Seasonal and daily spatio-temporal patterns may influence the intensity of interspecific interactions and the resulting distribution of sympatric species [34]. In a Scandinavian context, wolves prey on moose all year round, bears often scavenge wolf kills, and both predators largely rely on neonate moose during spring [20]. Thus, individual and seasonal variation in predation and scavenging rates may help understand the patterns of wolf habitat selection and kill rates in relation to brown bear presence reported earlier. Namely, bear density has had a negative effect on the probability of wolf territory establishment in Scandinavia during the wolf recolonization process [16,18], and wolf kill rates are lower in areas sympatric with bears, despite wolves losing food to bears via scavenging [20].

In this study, we characterized the temporal use of a common prey resource by sympatric wolves and bears in Sweden, describing individual and seasonal variation in wolf–bear interactions. The latter can be direct, if individual wolves and bears meet at the same time, or indirect, if they use the same place or resource, but not simultaneously, which can provide evidence of exploitation competition, e.g., via scavenging. Describing individual variation and seasonal trends in kill and scavenger rates of competing carnivores can reveal underlying mechanisms behind the observed effects of bears on wolf habitat selection and kill rates at higher spatial and temporal scales. Ecological theory is improving the forecast of changes in species interactions and coexistence in a scenario of global change, but more specific empirical data are needed to understand the mechanisms driving interactions and thus species coexistence (e.g., [29]). Our study contributes empirical data to document the role of individual variation and seasonality as drivers of interspecific interactions between apex predators via predation rates and scavenging, which in turn can also reflect on predator-prey interactions.

2. Material and Methods

2.1. Study Area and Study Species

The study area in central Sweden (in Dalarna, Jämtland, and Gävleborg counties) is a rolling landscape mainly covered by boreal, coniferous forests dominated by Scots pine *Pinus sylvestris* and Norway spruce *Picea abies*. Altitudes range between 100–830 m. Human density is low, 1–7 inhabitants/km^2, but the landscape is crisscrossed by many gravel roads (1 ± 0.5 km/km^2), because logging is a major activity [35]. Snow typically covers the ground from December to March. Wolves became functionally extinct in Scandinavia in the 1960s, but wolf recolonization and recovery started by the late 1970s [36] and continued until 2015, when the general increasing population trend stabilized [16,37]. The first wolf territories reestablished in our bear-wolf sympatric study area in 2000/2001, and afterwards between one and eight wolf territories have been detected annually [17], with a pack size of 4 ± 2 wolves (mean ± sd) for the packs included in our study. For brown bears, as few as about 130 were left in Sweden about a century ago [38], but human attitudes and legislation changed, the population recovered, and currently bear density reaches ~30 bears/1000 km^2 in our study area [39]. As of winter 2019–2020, there were ~450 wolves in Scandinavia, ~365 of them in Sweden [37]. The Scandinavian brown bear population consists of ~3000 bears, most of them (~2800) in Sweden [40]. Lynx and wolverines are also present, yet in low densities, in the study area. Moose is the most abundant ungulate (0.7–1.6/km^2) and very low densities of roe deer *Capreolus capreolus* (0.05–0.08/km^2) also occur [41].

2.2. GPS and Predation Data from Bears and Wolves

Studies of predation by wolves and bears on moose in the study area were conducted during two time periods; late winter and early spring (hereafter, "late winter"; from mid-February to the end of April) and early summer (hereafter, "early summer"; from the beginning of May to early July). For this study, we conducted predation studies of wolves in 2010–2015, and of bears overlapping with wolf territories in 2014 and 2015. Both wolves and bears were darted and immobilized from helicopters, according to accepted veterinary and ethical procedures [42], as determined by an ethical committee (Djurförsöksetisk nämnd) and the wildlife management authorities (Naturvårdsverket). The breeding pair in each wolf pack (6 wolves in total) was equipped with a GPS collar (Vectronic Aerospace, Berlin, Germany) and was monitored during each study period. In 2014, we monitored the predatory behavior of two wolf packs and 11 radio-collared bears with overlapping territories, and in 2015 we monitored one of the packs and nine collared bears (Tables 1 and 2; see also Figure 2 in [17]). There are more bears in the study area, but up to 80% of the adult female bears and 50% of the adult males have been radio-collared [43].

We searched for carcasses of killed prey at clusters of GPS locations [20,26] and recorded cause of death and age of the dead animal. We built clusters independently for bears and wolves, and we used the time of first bear or wolf location within each cluster of GPS locations as the time of death of each killed prey. We downloaded and plotted the GPS locations in ArcView GIS (Environmental Systems Research Institute, Inc., Redlands, CA, USA). We created a buffer around each location with a radius of 100 m and overlapping buffers generated clusters of ≥2 locations [44]. We uploaded them into handheld GPS receivers (Garmin, Olathe, KS, USA) and we visited all the generated clusters of GPS locations of wolves and bears within the study periods in the field. As in previous studies [20,26,45], we generally visited the clusters of locations after 3 days, trading off carcass detection and avoidance of disturbance of study animals. Kills were relatively easy to find; even the predation of a neonate moose removed the understory vegetation, as previously reported [26] and typically the jaws and other small pieces of bones and skin were present. As additional sign, within each cluster of GPS locations we also recorded if there were tracks, scats, and hair of wolves or bears in the ground and lying vegetation, at the sites where prey were consumed and/or at surrounding daybeds within the clusters.

Table 1. Individual variation in the number of neonate moose calves killed by brown bears in central Sweden in early summer, during the moose calving season, according to bear sex and age categories, including the specific period in which each bear killed moose calves.

Bear	Year	Predation Period		Predation Period (# Days)	# Killed Calves	Bear Sex and Age Class
Galju	2014	16.05.2014	17.06.2014	32	10	Single adult female
Ruta	2014	18.05.2014	18.06.2014	31	5	Single adult female
Spjuta	2014	21.05.2014	12.06.2014	22	10	Single adult female
Strandas	2014	03.06.2014	09.06.2014	6	2	Single adult female
Klummy	2014	15.06.2014	15.06.2014	1	1	Subadult female
	2015	08.05.2015	18.06.2015	41	4	Single adult female
Klumpa	2014	10.06.2014	25.06.2014	15	4	Subadult female
	2015	06.05.2015	19.05.2015	13	1	Single adult female
Jarpa	2014	–	–	0	0	Subadult female
	2015	–	–	0	0	Subadult female
Lafmamack	2014	23.05.2014	24.06.2014	32	3	Adult male
Risslo	2014	13.05.2014	06.06.2014	24	7	Adult male
	2015	20.05.2015	21.06.2015	32	4	Adult male
Tappele	2014	22.05.2014	01.06.2014	10	10	Adult male
	2015	11.05.2015	17.06.2015	37	4	Adult male
Kil-kalle	2014	28.05.2014	09.06.2014	12	3	Subadult male
	2015	28.05.2015	10.06.2015	13	3	Adult male
Lutane	2015	18.05.2015	21.06.2015	34	2	Adult male
Abborrgina	2015	23.05.2015	19.06.2015	27	8	Adult female + 2 year-old cubs
Gymasa	2015	18.05.2015	20.06.2015	33	4	Single adult female

Table 2. Wolf territories for which GPS clusters were visited in 2010-2015. Start date and end date columns denote the time during which clusters were checked. Fieldwork in 2014 and 2015 was extended during the moose calving season, thus two fieldwork seasons are differentiated (a and b) for the wolf territories tracked during those years. Number and age class of killed moose per wolf territory and study period is also reported.

Wolf Territory	Year	Start Date	End Date	Killed prey (# per Moose Age Class)
Tenskog	2010	2/13/2010	4/11/2010	5 juveniles
	2011	3/14/2011	5/16/2011	8 juveniles
Tandsjön	2012	2/20/2012	5/14/2012	1 neonate calf, 5 juveniles, 7 adults, 1 unknown (juvenile or adult)
	2014a	3/19/2014	4/28/2014	3 juveniles
	2014b	5/21/2014	6/21/2014	11 neonate calves, 3 juveniles
Kukumäki	2013	2/25/2013	4/28/2013	2 juveniles, 2 adults
	2014a	3/3/2014	4/27/2014	3 juveniles
	2014b	5/19/2014	6/22/2014	7 neonate calves, 6 juveniles
	2015a	3/4/2015	4/24/2015	11 juveniles, 1 unknown (juvenile or adult)
	2015b	5/18/2015	6/29/2015	16 neonate calves, 2 juveniles, 1 adult

2.3. Monitoring of Scavenging with Camera Traps

To record visits of wolves and bears at kills made by any of the predators, in 2014–2015 we placed camera traps (Reconyx, Holmen, WI, USA and Scoutguard, Santa Clara, CA, USA) targeting the carcasses and immediate surroundings. We placed cameras at all kill sites with some remaining biomass. Cameras were triggered by movement sensors from passing animals and were programed to record three pictures at a time, with a 1-min time lapse until the next three-picture set. We left the cameras in the field for at least 2 weeks or until the kill had been completely consumed (typically, no leftovers at all or only the jaws and/or some skin and hair of the moose were left).

2.4. Data Analyses

2.4.1. Kill Rates

We calculated average daily kill rates of moose by wolves (at the pack level) and bears during the study period. We estimated the variation in wolf and bear kill rates over time by calculating the average number of moose killed/day within a 7-day moving window. We then used bootstrapping to derive 95% confidence intervals. Wolves are obligate carnivores and mostly prey on moose in the study area [41], whereas bears are omnivores that use multiple food items, including moose calves in early summer [46]. Thus, we also checked for individual variation in the predation pattern of bears. In the early summer study period, i.e., the peak of the moose calving season, we summed the number of moose calves killed by each bear and calculated the average number of moose calves killed by the different sex and age categories of bears (<4 years old were considered as subadults [17]), to test for sex- and/or age-related differences in kill rates by using nonparametric Kruskal–Wallis and Mann–Whitney U-tests. The moose calving season overlaps the bear mating season (see Figure 3 in [17] for a graphical description of the seasonal phenology of bears, wolves, and moose). At that time, female bears with cubs of the year have small home ranges and limited movements to avoid conspecifics [47]. Therefore, we did not visit GPS clusters of females with cubs of the year in this study to prevent their displacement. However, we visited the clusters of one female with yearling cubs in 2015 (it was included as an "adult female" in the tests, i.e., we did not have a specific "female with yearlings" category because it was only one family group).

2.4.2. Scavenging Events

We defined a scavenging event as ≥1 picture of one of the large carnivores per day at a kill. We calculated the average number of scavenging events per day across all active cameras within a 7-day moving window and used bootstrapping to derive 95% confidence intervals.

3. Results

We visited a total of 1051 clusters of GPS locations of bears (530 clusters in 2014 and 521 in 2015) and 891 clusters of wolf GPS locations (97 in 2010, 169 in 2011, 179 in 2012, 76 in 2013, 218 in two territories in 2014, and 152 in 2015).

3.1. Direct Interactions Derived from GPS-Data and Field Visits

The only direct interaction (i.e., recorded GPS locations of both wolf and bear meeting simultaneously at the same place) that we could confirm occurred in 2012 at a yearling moose carcass killed by wolves in late March. The male wolf of the pair approached the carcass (25 days later) and stayed nearby for a day, but did not access it, likely due to the presence of a 9-year old male bear that was feeding at the carcass. There was likely another direct interaction in late June 2014, but the incomplete success of wolf GPS locations that day prevented confirmation. In the latter case, both bear and wolf locations overlapped on the remains of a neonate moose, which had presumably been killed by the bear, based on sign (a bear daybed and hair) found at the spot. Therefore, direct interactions with bears and wolves meeting simultaneously at the same place seem to be very rare in our study area.

3.2. Temporal Patterns in Bear and Wolf Kill Rates

In total, we found 85 moose neonate calves killed by 14 different bears during a total of 19 different bear-years (for 1 bear, we did not find any killed moose) (Tables 1 and A1), and 95 moose killed by wolves, including 35 neonate calves, 48 juvenile moose (<12 months old), 10 ≥ 1-year old, and 2 moose for which it could not be determined if they were yearlings or older individuals, in 7 different wolf territory-years, some of them with two predation studies (spring and summer) per year (Table 2). During the annual study periods between late February and early July, wolf pack kill rate averaged ~0.21 moose killed per day, yet it showed much variation and peaked at 0.65 moose killed per day in late May (Figure 1). Bears, with a shorter predation season, started to kill neonate moose calves around mid-May and stopped around the end of June, i.e., bear predation was limited to the moose calving season. Bear kill rates also peaked in late May, reaching a maximum of 0.4 moose killed per day and an average of 0.08 moose killed per day during the predatory period (Figure 1).

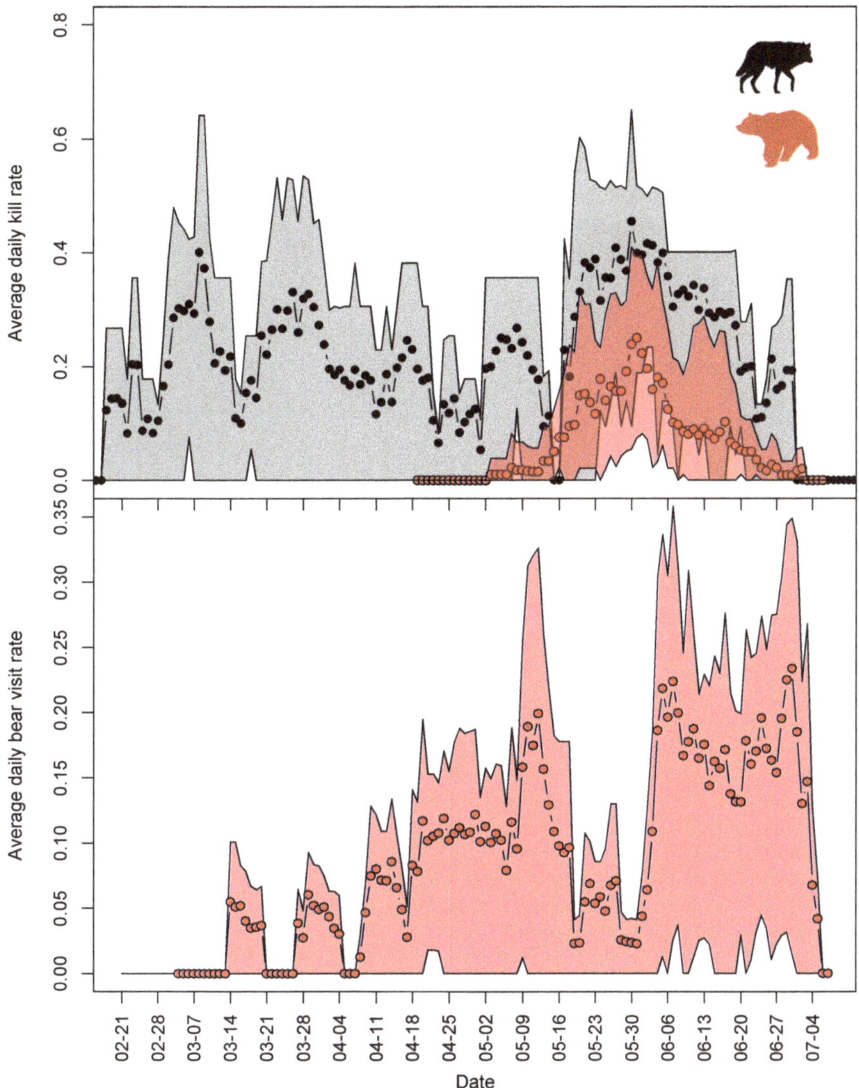

Figure 1. Average brown bear (red dots) and gray wolf (black dots) kill rates and 95% confidence intervals (gray and red shades) in 2014–2015 in central Sweden, calculated with a 7-day moving window (top panel). Daily average bear visits to wolf-killed moose carcasses and 95% confidence intervals in 2014–2015 in central Sweden, calculated with a 7-day moving window (bottom panel). In both panels, the *x*-axis displays the date.

On average, a bear killed 4.25 moose calves per season (sd = 3), but there was large individual variation. Whereas some bears killed up to 10 calves during an early-summer period, one did not kill any (Table 1). On average, an adult female bear killed 5.5 calves (sd = 3.46, $n = 8$), a subadult female killed 1.25 calves (sd = 1.9, $n = 4$), an adult male killed 4.71 calves (sd = 2.81, $n = 7$), and one subadult male killed three calves during the early-summer period overlapping the moose calving season. Adult bears killed significantly more moose calves than subadults (Mann-Whitney U = 63.5,

$p = 0.02$), with no significant differences in kills rates between male and female bears (Mann-Whitney U = 43.5, $p = 0.75$) or when combining sex and age classes (Kruskal–Wallis chi-squared = 5.74, $p = 0.12$; Figure 2). The bear predatory period in early summer, i.e., the time span between the first and last kill of neonate calves by bears in a given year, showed much individual variation; e.g., the average predatory period for bears that killed at least two calves in a season was 25 days (sd = 11, range 6–41; Table 1).

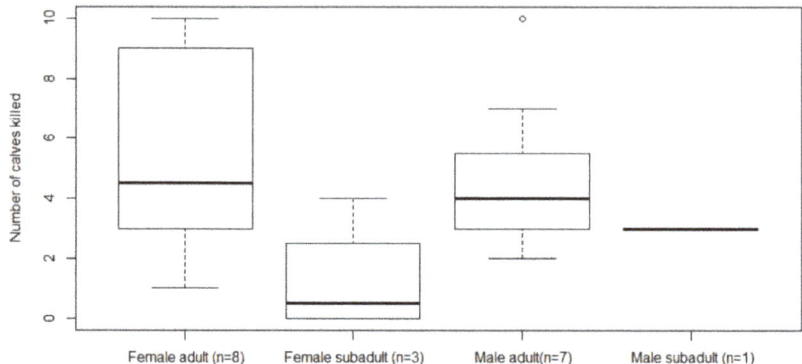

Figure 2. Adult brown bears killed significantly more moose calves in central Sweden than subadult bears, but there were no significant differences in kill rates between male and female bears or when combining sex and age classes, during the early summer period overlapping the moose calving season.

Bear predation focused exclusively on moose neonate calves, and wolves also preyed mostly upon neonate moose in the early-summer period. From mid-May onwards, 81% of the wolf kills were neonate calves and 16% were juvenile (~12 month old) moose, whereas earlier in the year, from late February until the moose calving period started in May, 81% of the wolf kills were juvenile moose <12 months and the rest, older individuals. Therefore, the moose calving period was the time when bears and wolves shared the same resource, with both species relying on neonate moose calves as main prey.

3.3. Scavenging Events

We placed cameras at 62 different kill sites in 2014–2015 (40 at wolf-killed carcasses, most of them yearling and adult moose, and 22 at bear-killed carcasses of neonate moose). Bears were seen on 58% of all cameras, which resulted in 3572 pictures and 122 different events. Wolves were detected on 27% of all cameras, which resulted in 366 pictures of wolves in 31 different events. Bear pictures (of both collared and uncollared bears) were taken at 50% (in 2014) and 60% (in 2015) of the wolf kills, but no wolf picture was taken at bear kills. Occurrence of bears at a carcass varied over time, with a lower average number of bear visits from mid-May to early June, the period when bear kill rates were highest (Figure 1). Adult and subadult bears visited wolf kills, as confirmed by photos of our collared bears, but we did not record any visit of a female bear with cubs of the year at wolf kills, and only two wolf kills were visited by uncollared bear females with 2-year-old cubs. Bears were photo-trapped at kill sites throughout the 24 h, with a peak in the number of bear pictures taken during the evening, whereas wolves were most often photographed at kills during late evening and, especially, at nighttime (Figure A1).

4. Discussion

Our study shows that indirect, nonsimultaneous interactions between wolves and bears at wolf kill sites are the norm, with direct interactions being rare. Bears used wolf kills very often, whereas wolves did not visit, potentially with one exception, bear kills. All neonate moose calves that bears killed were completely consumed shortly after the kill. However, two thirds of the moose killed by wolves (60 of 95) were ≥9–10 months old; thus, the large carcasses provided feeding opportunities to scavengers. Some individual bears were efficient predators, but preyed exclusively on neonate moose, as described earlier in our study system [26,48] and elsewhere [15]. Bears are also efficient scavengers of other predators' kills [49], as confirmed in our study, with ≥50% of the wolf kills visited and thus presumably scavenged by brown bears each study year.

The start of the moose calving period, around mid-May, triggered a clear change in the behavior of some individual bears, which became predatory, but not all did. The average number of neonate moose calves killed by bears during the early-summer study period (4.25) seemed lower than estimated in earlier studies in the same area (6.8 calves for bears ≥4 year old in Swenson et al. 2007 [48], and 7.6 calves for 3–13 year old female bears in Rauset et al. 2012 [26]). However, moose densities have declined there (and elsewhere in Sweden) from the 1990s, possibly resulting in a functional response by bears, and in any case preventing direct comparisons [26]. Furthermore, the number of wolf territories in the study area increased substantially after the earlier studies on bear predation were carried out [16], likely causing further changes in the moose population and increasing the number of wolf-killed prey to be eventually scavenged by bears. Most importantly, earlier studies on bears also reported large individual variation in per capita kill rates, 2 to 15 calves per season, which was attributed to a large individual variation in hunting skills and maybe effort [26]. In our study, several single female and male adult bears were the most predatory, preying upon calves in a period up to 4–5 weeks, whereas younger bears killed fewer or even no calves (Table 1). Thus, it seems that some bears actively searched for moose neonates during the calving season, whereas others killed fewer or none moose calves, denoting a more opportunistic predatory behavior and/or different levels of hunting experience and ability. Individual- and age-class differences may reduce intraspecific competition among bears and, most important in the context of this study, may promote interspecific coexistence between wolves and bears [17]. Individual foraging specialization, i.e., different predatory levels among individual bears and bear age classes, could be one of the mechanisms involved reducing competition with the (obligate carnivorous) wolves.

Besides the fact that direct wolf–bear interactions were very rare, no wolves were photographed at bear kills, no females bears with cubs of the year were photographed at wolf kills, and only two females with 2-year-old cubs were photographed at two wolf kills. However, visits of single bears (adults and subadults) at wolf kills were common. Photo-trapped bears at kills were typically smelling and/or biting and moving the carcasses, either standing or lying on top of them, thus reflecting that they were scavenging on them. Wolf absence at bear kills (neonate moose) is likely explained by the fact that little or no biomass is left by the bear. We could not quantify the relative frequency of bear use of wolf kills by bear age and sex class, because assigning bear sex and age class to uncollared bears, which often visited kills based on photos taken by camera traps, is not reliable. However, an interesting result was that females with cubs of the year did not use wolf kills, despite a minimum of 19 collared females with offspring partially overlapped the wolf territories included in our study in 2014 and 2015 (Scandinavian Brown Bear Research Project monitoring data). Bear females with cubs avoid conspecifics spatially and temporally [47,50] and both bears and wolves can kill the offspring of each other [12]. Single bears were photographed at kills throughout the day (Figure A1), likely explaining why we did not record any visit of females with cubs at wolf-killed carcasses. Altogether, different levels of trophic specialization and fine-scale spatial avoidance, with virtually no bear-wolf direct interactions and no use of wolf kills by the most vulnerable bear classes, are mechanisms reducing predation risk and favoring coexistence between sympatric large carnivores.

Fine-scale spatial segregation among large carnivores can be achieved via fine-scale movement patterns, with neighboring individuals avoiding each other when they use shared resources at kill sites [51]. In turn, fine-scale spatial segregation may be a mechanism reflecting on individual habitat selection at larger scales. We have earlier shown that bear density has had a negative effect on the probability of wolf territory establishment during the wolf recolonization of central Scandinavia [16,18] and that overlapping wolves and bears use different habitat types to a larger extent than expected [17]. Wolves seem to avoid bears at different spatial scales, yet wolf habitat selection within home ranges is not different in areas sympatric and allopatric with bears [19]. The results of our present study show that bears, except females with offspring, take advantage of wolf predation; i.e., indirect interactions at wolf kills sites where bears feed, are hotspots mediating coexistence between these species.

We visited carcasses typically 3 days after a prey was killed to avoid disturbance; human scent may influence the frequency and timing of subsequent visits by the predators, and individual bears and wolves may display different levels of reaction to such disturbance. Most often, >80–90% of the wolf kills (and ~100% of the bear kills) had been already consumed at our first visit. If the wolves returned to a previous kill to resume feeding and it had been partially depleted by bears and/or other scavengers, kleptoparasitism is the ongoing process. If, however, bears and other scavengers use already-abandoned wolf kills, facilitation would be the dominant mechanism. We recorded 31 visits of wolves at ~ one third (27%) of their kills, i.e., wolves returned to their kills presumably to resume feeding. The level of consumption of wolf kills, as visually determined at our first visit, is a rough proxy of available biomass at the carcass and it is very similar in bear and non-bear areas (authors' unpublished data). Thus, bear consumption of wolf kills may not necessarily impact wolves in terms of food loss, because wolf kill rates are not higher in bear areas than in non-bear areas [20]. Both kleptoparasitism by bears, which are able to remove substantial amounts of the carcass, and facilitation seem to occur in our system, and these mechanisms likely occur elsewhere, over the vast range where these species overlap.

Direct interactions between wolves and bears might occur at sites other than carcasses, such as breeding dens, where consequences could be most dramatic, e.g., in terms of offspring survival. Nevertheless, all clusters of GPS-locations we have identified and visited in the field (n = 1942 clusters), were either daily resting sites (daybeds) and places with no observable sign (90% of the clusters) or sites with predated moose (10%). Even though our predation studies in spring overlapped the breeding season of wolves (and the monitored wolf packs reproduced during the study period) and the season when female bears are coming out of winter dens with their newly born cubs, we did not record any wolf or bear predatory attempt on offspring of the other species. These results reinforce that the vast majority of interactions between wolves and bears occur at kill sites [12], and help discuss mechanisms regulating coexistence of these apex predators.

In Scandinavia, as in other boreal ecosystems, bears rely mainly on berries during summer and autumn, before winter denning [46], and they are also efficient scavengers [49]. Our study documents that during early summer, bears reduce scavenging and actively start to prey on the very same resource that wolves exploit. This defines a seasonal gradient of interspecific competition that peaks in early summer, when wolves also switched from predation on juvenile moose (~80% of wolf kills in late winter) to neonate moose (~80% of wolf kills in early summer). These seasonal prey age preferences by Scandinavian wolves reinforce previous findings [45], and our study documents the importance of both individual variation (in bear predation) and seasonality (in bear diet, predation and scavenging patterns, and in wolf predation on varying prey age classes) for understanding interspecific interactions [33,34].

The role of individual variation, sometimes referred to as personality, at the population level is increasingly recognized in ecology [31,52,53]. Some bears use specific food items, e.g., of anthropogenic origin, that other bears do not use and that reflects on individual differences in movements patterns and habitat selection [54]. The variation in bear predatory behavior at individual and age-class levels documented in our study builds upon this topic. The predatory behavior of a bear may depend on social learning during the ≥1.5 year spent with its mother, as suggested for sea otters (*Enhydra lutris*). Consistent individual variation in sea otters' diet could be related to the matrilineal transmission of foraging preferences and/or skills [55]. Matrilines have been documented in Scandinavian brown bears [56], which makes this species suitable for further studies on the topic of individual variation, cultural transmission across generations, and the implications it can have in different grounds. For instance, removal of bears and other large carnivores is a management tool to reduce depredation on livestock, in Scandinavia [15] and elsewhere [4], thus the individual identification and eventual removal of most predatory individuals, rather than indiscriminate removals, could increase management efficiency. Individual variation in large carnivore behavior is indeed gaining recognition in conservation-oriented research [57]. We suggest that individual variation is also important to understand the outcome of interspecific interactions at higher levels of biological organization, i.e., at the population level of involved species.

Interspecific interactions between species of different size (e.g., larger predators controlling mesopredators and prey numbers [4,58]) and interactions in carnivore assemblages that have been coexisting for a long time, such of those in some African ecosystems (e.g., [59,60]), are better documented than interspecific interactions between top predators, especially in recolonizing scenarios in the northern hemisphere. The recent recovery of some large carnivore populations in North America and Europe allow for overlapping distribution of competing species, and for studies on interspecific interactions and their effects on predation and scavenging patterns [20,61]. Long-term monitoring of bears and wolves combined with intensive fieldwork help understand the mechanisms involved in the coexistence of these apex predators. Therefore, we suggest that this approach can shed light to similar processes elsewhere, in ecosystems holding the same and/or alternative species assemblages.

Author Contributions: Conceptualization, C.M. and A.O.; Methodology, C.M., A.O., and A.U.; Formal Analysis, C.M., A.O., and A.U.; Investigation and Data Curation, A.O. and C.M.; Resources, B.Z., P.W., C.W., H.S., J.E.S., and J.K.; Writing—Original Draft Preparation, A.O. and C.M.; Writing—Review & Editing, all authors; Visualization, C.M. and A.O.; Supervision and Project Administration, B.Z., P.W., C.W., H.S., J.E.S., and J.K.; Funding Acquisition, A.U., B.Z, P.W., C.W., H.S., J.E.S., and J.K. All authors have read and agreed to the submission of the manuscript. All authors have read and agreed to the published version of the manuscript.

Funding: The Scandinavian wolf and bear research projects (SKANDULV and SBBRP, respectively) have been funded by the Norwegian Environment Agency, Swedish Environmental Protection Agency, Norwegian Research Council, Swedish Research Council Formas, Austrian Science Fund, Norwegian Institute for Nature Research, Inland Norway University of Applied Sciences, Swedish University of Agricultural Sciences, Office of Environmental Affairs in Hedmark County, Swedish Association for Hunting and Wildlife Management, WWF (Sweden), Swedish Carnivore Association, Olle and Signhild Engkvists Foundations, Carl Tryggers Foundation, and Marie-Claire Cronstedts Foundation. A.O. and C.M. received QR funding from Nottingham Trent University.

Acknowledgments: We are grateful to everyone monitoring wolves and bears in Scandinavia, including the teams that captured and handled the animals. E.R. Dahl, F. Holen, K. Nordli, J. Romairone, M. Rostad, D. Roviani, and A. Tallian helped us in the field in 2014–2015. GPS data was collected into the Wireless Remote Animal Monitoring database system for data validation and management [62]. We acknowledge the editor and reviewers for constructive comments during the review process.

Appendix A

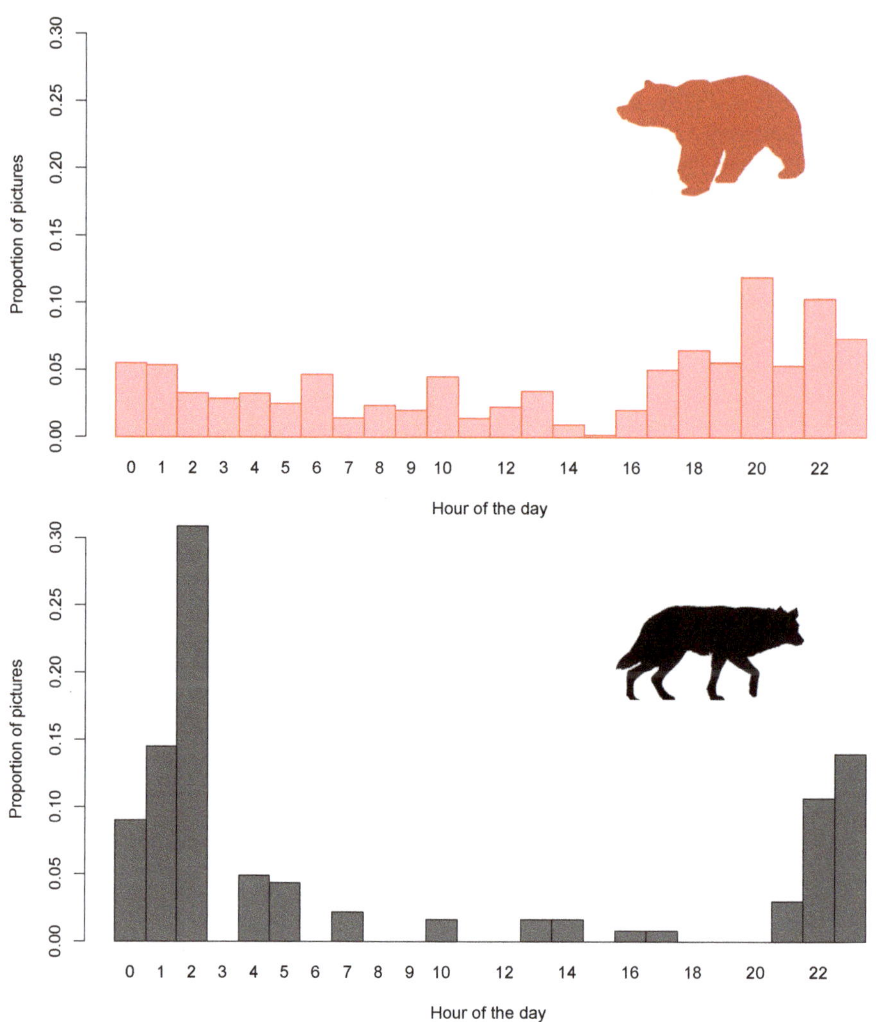

Figure A1. Proportion of bear and wolf pictures photo-trapped per time of day at kill sites (mostly, at wolf kill sites) in central Sweden.

Table A1. Brown bears for which GPS-clusters were visited. Start date and end date columns denote the time during which clusters were checked, i.e., when bears were inside the study area overlapping wolf territories. The study period overlapped the bear mating season, so some males occasionally left the study area and returned. We visited their clusters when they were inside the study area.

Bear and Year of Study	Start Date	End Date
Lafmamack 2014	5/14/2014	5/14/2014
Lafmamack 2014	5/22/2014	5/27/2014
Lafmamack 2014	6/3/2014	6/9/2014

Table A1. *Cont.*

Bear and Year of Study	Start Date	End Date
Lafmamack 2014	6/12/2014	6/27/2014
Galju 2014	4/28/2014	7/1/2014
Klummy 2014	4/30/2014	5/9/2014
Klummy 2014	5/21/2014	5/26/2014
Klummy 2014	6/12/2014	6/19/2014
Klummy 2014	6/23/2014	6/27/2014
Klumpa 2014	4/30/2014	5/9/2014
Klumpa 2014	5/21/2014	6/27/2014
Risslo 2014	4/22/2014	6/23/2014
Ruta 2014	4/29/2014	6/27/2014
Jarpa 2014	5/6/2014	6/23/2014
Kil-kalle 2014	4/30/2014	6/27/2014
Spjuta 2014	5/11/2014	7/1/2014
Strandas 2014	5/11/2014	6/27/2014
Tappele 2014	5/11/2014	7/1/2014
Abborrgina 2015	5/3/2015	7/1/2015
Gymasa 2015	5/13/2015	7/3/2015
Jarpa 2015	6/12/2015	7/3/2015
Kil-kalle 2015	4/27/2015	5/3/2015
Kil-kalle 2015	5/13/2015	5/16/2015
Kil-kalle 2015	5/27/2015	5/30/2015
Kil-kalle 2015	6/6/2015	6/15/2015
Kil-kalle 2015	6/24/2015	7/3/2015
Klummy 2015	4/27/2015	7/3/2015
Klumpa 2015	4/27/2015	5/25/2015
Klumpa 2015	6/6/2015	6/9/2015
Klumpa 2015	6/21/2015	6/24/2015
Lutane 2015	4/27/2015	5/31/2015
Lutane 2015	6/6/2015	7/3/2015
Risslo 2015	4/27/2015	5/22/2015
Risslo 2015	6/9/2015	6/15/2015
Risslo 2015	6/21/2015	7/3/2015
Tappele 2015	4/27/2015	6/24/2015

References

1. Terborgh, J.; Estes, J.A. *Trophic Cascades*; Island Press: Washington, DC, USA, 2010.
2. Ripple, W.J.; Smith, D.W.; Estes, J.A.; Beschta, R.L.; Wilmers, C.C.; Ritchie, E.G.; Hebblewhite, M.; Berger, J.; Elmhagen, B.; Letnic, M.; et al. Status and ecological effects of the world's largest carnivores. *Science* **2014**, *343*, 1241484. [CrossRef] [PubMed]
3. Hebblewhite, M.; Hurd, T.E.; White, C.A.; Fryxell, J.M.; Nietvelt, C.G.; McKenzie, J.A.; Bayley, S.E.; Paquet, P.C. Human Activity Mediates a Trophic Cascade Caused By Wolves. *Ecology* **2005**, *86*, 2135–2144. [CrossRef]
4. Ordiz, A.; Bischof, R.; Swenson, J.E. Saving large carnivores, but losing the apex predator? *Biol. Conserv.* **2013**, *168*, 128–133. [CrossRef]
5. Bruskotter, J.T.; Shelby, L.B. Human Dimensions of Large Carnivore Conservation and Management. *Hum. Dimens. Wildl.* **2010**, *15*, 311–314. [CrossRef]
6. Chapron, G.; Kaczensky, P.; Linnell, J.D.C.; von Arx, M.; Huber, H.; Andrén, H. Recovery of large carnivores in Europe's modern human-dominated landscapes. *Science* **2014**, *346*, 1517–1519. [CrossRef]
7. May, R.; Van Dijk, J.; Wabakken, P.; Swenson, J.E.; Linnell, J.D.C.; Odden, J.; Pedersen, H.C.; Andersen, R.; Landa, A. Habitat differentiation within the large-carnivore community of Norway's multiple-use landscapes. *J. Appl. Ecol.* **2008**, *45*, 1382–1391. [CrossRef]
8. Wikenros, C.; Liberg, O.; Sand, H.; Andrén, H. Competition between recolonizing wolves and resident lynx in Sweden. *Can. J. Zool.* **2010**, *88*, 271–279. [CrossRef]
9. Mattisson, J.; Andrén, H.; Persson, J.; Segerström, P. Influence of intraguild interactions on resource use by wolverines and Eurasian lynx. *J. Mammal.* **2011**, *92*, 1321–1330. [CrossRef]

10. Elmhagen, B.; Ludwig, G.; Rushton, S.P.; Helle, P.; Lindén, H. Top predators, mesopredators and their prey: Interference ecosystems along bioclimatic productivity gradients. *J. Anim. Ecol.* **2010**, *79*, 785–794. [CrossRef]

11. Gervasi, V.; Nilsen, E.B.; Sand, H.; Panzacchi, M.; Rauset, G.R.; Pedersen, H.C.; Kindberg, J.; Wabakken, P.; Zimmermann, B.; Odden, J.; et al. Predicting the potential demographic impact of predators on their prey: A comparative analysis of two carnivore-ungulate systems in Scandinavia. *J. Anim. Ecol.* **2012**, *81*, 443–454. [CrossRef]

12. Ballard, W.; Carbyn, L.N.; Smith, D.W. Wolf interactions with non-prey. In *Wolves: Behavior, Ecology and Conservation*; Mech, L.D., Boitani, L., Eds.; University of Chicago Press: Chicago, IL, USA, 2003; pp. 259–271.

13. Gunther, K.A.; Smith, D.W. Interactions between wolves and female grizzly bears with cubs in Yellowstone National Park. *Ursus* **2004**, *15*, 232–238. [CrossRef]

14. Lewis, T.M.; Lafferty, D.J.R. Brown bears and wolves scavenge humpback whale carcass in Alaska. *Ursus* **2014**, *25*, 8–13. [CrossRef]

15. Ordiz, A.; Krofel, M.; Milleret, C.; Seryodkin, I.V.; Tallian, A.; Støen, O.-G.; Sivertsen, T.R.; Kindberg, J.; Wabakken, P.; Sand, H.; et al. Interspecific interactions between brown bears, ungulates, and other large carnivores. In *Bears of the World*; Penteriani, V., Melletti, M., Eds.; Cambridge University Press: Cambridge, UK, 2020; in press.

16. Ordiz, A.; Milleret, C.; Kindberg, J.; Mansson, J.; Wabakken, P.; Swenson, J.E.; Sand, H. Wolves, people, and brown bears influence the expansion of the recolonizing Wolf population in Scandinavia. *Ecosphere* **2015**, *6*, 1–14. [CrossRef]

17. Milleret, C.; Ordiz, A.; Chapron, G.; Andreassen, H.P.; Kindberg, J.; Månsson, J.; Tallian, A.; Wabakken, P.; Wikenros, C.; Zimmermann, B.; et al. Habitat segregation between brown bears and gray wolves in a human-dominated landscape. *Ecol. Evol.* **2018**, *8*, 1–17. [CrossRef] [PubMed]

18. Sanz-Pérez, A.; Ordiz, A.; Sand, H.; Swenson, J.E.; Wabakken, P.; Wikenros, C.; Zimmermann, B.; Akesson, M.; Milleret, C. No place like home? A test of the natal habitat-biased dispersal hypothesis in Scandinavian wolves. *R. Soc. Open Sci.* **2018**, *5*, 181379. [CrossRef] [PubMed]

19. Ordiz, A.; Uzal, A.; Milleret, C.; Sanz-pérez, A.; Zimmermann, B.; Wikenros, C.; Wabakken, P.; Kindberg, J.; Swenson, J.E.; Sand, H. Wolf habitat selection when sympatric or allopatric with brown bears in Scandinavia. *Sci. Rep.* **2020**, *10*, 1–11. [CrossRef]

20. Tallian, A.; Ordiz, A.; Metz, M.C.; Milleret, C.; Wikenros, C.; Smith, D.W.; Stahler, D.R.; Kindberg, J.; Macnulty, D.R.; Wabakken, P.; et al. Competition between apex predators? Brown bears decrease wolf kill rate on two continents. *Proc. R. Soc. B Biol. Sci.* **2017**, *284*, 20162368. [CrossRef]

21. Boertje, R.D.; Keech, M.A.; Paragi, T.F. Science and Values Influencing Predator Control for Alaska Moose Management. *J. Wildl. Manag.* **2010**, *74*, 917–928. [CrossRef]

22. Jonzén, N.; Sand, H.; Wabakken, P.; Swenson, J.E.; Kindberg, J.; Liberg, O.; Chapron, G. Sharing the bounty—Adjusting harvest to predator return in the Scandinavian human–wolf–bear–moose system. *Ecol. Model.* **2013**, *265*, 140–148. [CrossRef]

23. Rauset, G.R.; Mattisson, J.; Andrén, H.; Chapron, G.; Persson, J. When species' ranges meet: Assessing differences in habitat selection between sympatric large carnivores. *Oecologia* **2013**, *172*, 701–711. [CrossRef]

24. Mattisson, J.; Sand, H.; Wabakken, P.; Gervasi, V.; Liberg, O.; Linnell, J.D.C.; Rauset, G.R.; Pedersen, H.C. Home range size variation in a recovering wolf population: Evaluating the effect of environmental, demographic, and social factors. *Oecologia* **2013**, *173*, 813–825. [CrossRef] [PubMed]

25. Leclerc, M.; Vander, E.; Zedrosser, A.; Swenson, J.E.; Kindberg, J.; Pelletier, F. Quantifying consistent individual differences in habitat selection. *Oecologia* **2016**, *180*, 697–705. [CrossRef] [PubMed]

26. Rauset, G.R.; Kindberg, J.; Swenson, J.E. Modeling female brown bear kill rates on moose calves using global positioning satellite data. *J. Wildl. Manag.* **2012**, *76*, 1597–1606. [CrossRef]

27. Toscano, B.J.; Gownaris, N.J.; Heerhartz, S.M. Personality, foraging behavior and specialization: Integrating behavioral and food web ecology at the individual level. *Oecologia* **2016**, *182*, 55–69. [CrossRef]

28. Hertel, A.G.; Swenson, J.E.; Bischof, R. A case for considering individual variation in diel activity patterns. *Behav. Ecol.* **2017**, *28*, 1524–1531. [CrossRef]

29. Di Rienzo, N.; Montiglio, P. Four ways in which data-free papers on animal personality fail to be impactful. *Front. Ecol. Evol.* **2015**, *3*, 1–5. [CrossRef]

30. Davis, G.H.; Payne, E.; Sih, A. Commentary: Four ways in which data-free papers on animal personality fail to be impactful. *Front. Ecol. Evol.* **2015**, *3*, 102–104. [CrossRef]

31. Bolnick, D.I.; Amarasekare, P.; Araújo, M.S.; Bürger, R.; Levine, J.M.; Novak, M.; Rudolf, V.H.W.; Schreiber, S.J.; Mark, C. Why intraspecific trait variation matters in community ecology. *Trends Ecol. Evol.* **2011**, *26*, 183–192. [CrossRef]

32. Valladares, F.; Bastias, C.; Godoy, O.; Granda, E.; Escudero, A. Species coexistence in a changing world. *Front. Plant Sci.* **2015**, *6*, 1–16. [CrossRef]

33. Basille, M.; Fortin, D.; Dussault, C.; Ouellet, J.P.; Courtois, R. Ecologically based definition of seasons clarifies predator-prey interactions. *Ecography* **2013**, *36*, 220–229. [CrossRef]

34. Araújo, M.B.; Rozenfeld, A. The geographic scaling of biotic interactions. *Ecography* **2014**, *37*, 406–415. [CrossRef]

35. Ordiz, A.; Kindberg, J.; Sæbø, S.; Swenson, J.E.; Støen, O.G. Brown bear circadian behavior reveals human environmental encroachment. *Biol. Conserv.* **2014**, *173*, 1–9. [CrossRef]

36. Wabakken, P.; Sand, H.; Liberg, O.; Bjärvall, A. The recovery, distribution, and population dynamics of wolves on the Scandinavian peninsula, 1978–1998. *Can. J. Zool.* **2001**, *725*, 710–725. [CrossRef]

37. Wabakken, P.; Svensson, L.; Maartmann, E.; Nordli, K.; Flagstad, Ø.; Åkesson, M. *Bestandsovervåking av ulv vinteren 2019-2020 (In Norwegian)*; Rovdata and Viltskadecenter; SLU: St. Louis, MO, USA, 2020.

38. Swenson, J.E.; Wabakken, P.; Sandegren, F.; Bjärvall, A.; Franzén, R.; Söderberg, A. The near extinction and recovery of brown bears in Scandinavia in relation to the bear management policies of Norway and Sweden. *Wildl. Biol.* **1995**, *1*, 11–25. [CrossRef]

39. Solberg, K.H.; Bellemain, E.; Drageset, O.M.; Taberlet, P.; Swenson, J.E. An evaluation of field and non-invasive genetic methods to estimate brown bear (Ursus arctos) population size. *Biol. Conserv.* **2006**, *128*, 158–168. [CrossRef]

40. Swenson, J.E.; Schneider, M.; Zedrosser, A.; Söderberg, A.; Franzén, R.; Kindberg, J. Challenges of managing a European brown bear population; lessons from Sweden, 1943–2013. *Wildl. Biol.* **2017**, *1*, wlb.00251. [CrossRef]

41. Sand, H.; Eklund, A.; Zimmermann, B.; Wikenros, C.; Wabakken, P. Prey selection of Scandinavian wolves: Single large or several small? *PLoS ONE* **2016**, *11*, e0168062. [CrossRef]

42. Arnemo, J.; Fahlman, Å.; Ahlqvist, P.; Andersen, R.; Andrén, H.; Brunberg, S.; Landa, A.; Liberg, O.; Odden, J.; Persson, J.; et al. *Biomedical Protocols for Free-Ranging Brown Bears, Gray Wolves, Wolverines and Lynx*; Arnemo, J., Fahlman, A., Eds.; Norwegian School of Veterinary Science: Tromsø, Norway, 2007.

43. Ordiz, A.; Støen, O.-G.; Swenson, J.E.; Kojola, I.; Bischof, R. Distance-dependent effect of the nearest neighbor: Spatiotemporal patterns in brown bear reproduction. *Ecology* **2008**, *89*. [CrossRef]

44. Sand, H.; Zimmermann, B.; Wabakken, P.; Andrèn, H.; Pedersen, H.C. Using GPS technology and GIS cluster analyses to estimate kill rates in wolf—ungulate ecosystems. *Wildl. Soc. Bull.* **2005**, *33*, 914–925. [CrossRef]

45. Sand, H.; Wabakken, P.; Zimmermann, B.; Johansson, Ö.; Pedersen, H.C.; Liberg, O. Summer kill rates and predation pattern in a wolf-moose system: Can we rely on winter estimates? *Oecologia* **2008**, *156*, 53–64. [CrossRef]

46. Stenset, N.E.; Lutnæs, P.N.; Bjarnadóttir, V.; Dahle, B.; Fossum, K.H.; Jigsved, P.; Johansen, T.; Neumann, W.; Opseth, O.; Rønning, O.; et al. Seasonal and annual variation in the diet of brown bears *Ursus arctos* in the boreal forest of southcentral Sweden. *Wildl. Biol.* **2016**, *22*, 107–116. [CrossRef]

47. Ordiz, A.; Rodríguez, C.; Naves, J.; Fernández, A.; Huber, D.; Kaczensky, P.; Mertens, A.; Mertzanis, Y.; Mustoni, A.; Palazón, S.; et al. Distance-based criteria to identify minimum number of brown bear females with cubs in Europe. *Ursus* **2007**, *18*, 158–167. [CrossRef]

48. Swenson, J.; Dahle, B.; Busk, H.; Opseth, O.; Johansen, T.; Soderberg, A.; Wallin, K.; Cederlund, G. Predation on Moose Calves by European Brown Bears. *J. Wildl. Manag.* **2007**, *71*, 1993–1997. [CrossRef]

49. Krofel, M.; Kos, I.; Jerina, K. The noble cats and the big bad scavengers: Effects of dominant scavengers on solitary predators. *Behav. Ecol. Sociobiol.* **2012**, *66*, 1297–1304. [CrossRef]

50. Steyaert, S.M.J.G.; Kindberg, J.; Swenson, J.E.; Zedrosser, A. Male reproductive strategy explains spatiotemporal segregation in brown bears. *J. Anim. Ecol.* **2013**, *82*, 836–845. [CrossRef]

51. López-Bao, J.V.; Mattisson, J.; Persson, J.; Aronsson, M.; Andrén, H. Tracking neighbours promotes the coexistence of large carnivores. *Sci. Rep.* **2016**, *6*, 23198. [CrossRef]

52. Réale, D.; Reader, S.; Sol, D.; McDougall, P.; Dingemanse, N. Integrating animal temperament within ecology and evolution. *Biol. Rev. Camb. Philos. Soc.* **2007**, *82*, 291–318. [CrossRef]

53. Bergman, J.N.; Bennett, J.R.; Binley, A.D.; Cooke, S.J.; Fyson, V.; Hlina, B.L.; Reid, C.H.; Vala, M.A.; Madliger, C.L. Scaling from individual physiological measures to population-level demographic change: Case studies and future directions for conservation management. *Biol. Conserv.* **2019**, *238*, 108242. [CrossRef]

54. Cozzi, G.; Chynoweth, M.; Kusak, J.; Coban, E.; Coban, A.; Ozgul, A.; Sekerciouglu, C.H. Anthropogenic food resources foster the coexistence of distinct life history strategies: Year-round sedentary and migratory brown bears. *J. Zool.* **2016**, *300*, 142–150. [CrossRef]

55. Estes, J.; Riedman, M.; Staedler, M.; Tinker, M.; Lyon, B. Individual variation in prey selection by sea otters: Patterns, causes and implications. *J. Anim. Ecol.* **2003**, *72*, 144–155. [CrossRef]

56. Støen, O.-G.; Bellemain, E.; Solve, S.; Swenson, J. Kin-related spatial structure in brown bears Ursus arctos. *Behav. Ecol.* **2005**, *59*, 191–197. [CrossRef]

57. Suraci, J.P.; Nickel, B.A.; Wilmers, C.C. Fine-scale movement decisions by a large carnivore inform conservation planning in human-dominated landscapes. *Landsc. Ecol.* **2020**, *35*, 1635–1649. [CrossRef]

58. Palomares, F.; Caro, T.M. Interspecific Killing among Mammalian Carnivores. *Am. Nat.* **1999**, *153*, 492–508. [CrossRef] [PubMed]

59. Caro, T.M.; Stoner, C.J. The potential for interspecific competition among African carnivores. *Biol. Conserv.* **2003**, *110*, 67–75. [CrossRef]

60. Périquet, S.; Fritz, H.; Revilla, E. The lion king and the hyaena queen: Large carnivore interactions and coexistence. *Biol. Rev.* **2015**, *90*, 1197–1214. [CrossRef]

61. Wilmers, C.; Post, E. Predicting the influence of wolf-provided carrion on scavenger community dynamics under climate change scenarios. *Glob. Chang. Biol.* **2006**, *12*, 403–409. [CrossRef]

62. Dettki, H.; Ericsson, G.; Giles, T.; Norrsken-Ericsson, M. Wireless Remote Animal Monitoring (WRAM)—A new international database e-infrastructure for telemetry sensor data from fish and wildlife. In *Proceedings Etc 2012: Convention for Telemetry, Test Instrumentation and Telecontrol*; European Society of Telemetry; Books on Demand: Manila, Philippines, 2013; pp. 247–256. ISBN 978-3-7322-5646-4.

Article

Caching Behavior of Large Prey by Eurasian Lynx: Quantifying the Anti-Scavenging Benefits

Ivonne J. M. Teurlings [1,†], John Odden [2], John D. C. Linnell [1] and Claudia Melis [3,*]

1   Norwegian Institute for Nature Research, NO-7485 Trondheim, Norway; i_teurlings@hotmail.com (I.J.M.T.); john.linnell@nina.no (J.D.C.L.)
2   Norwegian Institute for Nature Research, Sognsveien 68, NO-0855 Oslo, Norway; john.odden@nina.no
3   Queen Maud University College for Early Childhood Education, Thrond Nergaards veg 7, 7044 Trondheim, Norway
*   Correspondence: cme@dmmh.no
†   Present address: Saghaugen gård, Skjomdalveien-Bogholm 21, 8523 Skjomen, Norway.

Received: 27 July 2020; Accepted: 10 September 2020; Published: 13 September 2020

Abstract: Large solitary felids often kill large prey items that can provide multiple meals. However, being able to utilize these multiple meals requires that they can cache the meat in a manner that delays its discovery by vertebrate and invertebrate scavengers. Covering the kill with vegetation and snow is a commonly observed strategy among felids. This study investigates the utility of this strategy using observational data from Eurasian lynx (*Lynx lynx*)-killed roe deer (*Capreolus capreolus*) carcasses, and a set of two experiments focused on vertebrate and invertebrate scavengers, respectively. Lynx-killed roe deer that were covered by snow or vegetation were less likely to have been visited by scavengers. Experimentally-deployed video-monitored roe deer carcasses had significantly longer time prior to discovery by avian scavengers when covered with vegetation. Carcass parts placed in cages that excluded vertebrate scavengers had delayed invertebrate activity when covered with vegetation. All three datasets indicated that covering a kill was a successful caching/anti-scavenger strategy. These results can help explain why lynx functional responses reach plateaus at relatively low kill rates. The success of this anti-scavenging behavior therefore has clear effects on the dynamics of a predator–prey system.

Keywords: caching; *Capreolus capreolus*; carrion; Eurasian lynx; *Lynx lynx*; Norway; predation; roe deer; scavenging

---

1. Introduction

The impacts of large carnivore predation on wild ungulate populations has received much research focus and considerable effort has been spent on quantifying predation rates as a function of prey density (i.e., functional responses, [1]) as these are regarded as key parameters to model predator impacts [2,3]. While functional responses may be crucial for some solitary species feeding on small prey and for social carnivores where group size allows rapid consumption of even large prey, there may be some predator–prey systems where other predation parameters are equally important. For example, large solitary felids routinely kill large ungulate prey equal to, or several times heavier than, their body weight that can potentially provide food for multiple days [4–7]. For such species handling and consumption time are likely to exceed search and killing time [8,9]. Any factor affecting their ability to consume a kill completely could result in an increased kill rate.

An ungulate carcass represents a very attractive resource to a range of vertebrate and invertebrate scavengers [10–15]. The effect of such scavenging and kleptoparasitic loss of kills is believed to be an important driver of kill rates [16], energetics [17,18], and potentially even sociality [19]. The issue has been widely explored for group hunting canid species, and has been reported for solitary felid species such as cheetah (*Acinonyx jubatus*) [20] and mountain lion (*Puma concolor*) [21]. Solitary felids are believed to use a range of behaviors, including covering the kill with snow and/or vegetation as an anti-scavenging/anti-kleptoparasitic caching strategy [22,23]. In the Bavarian forest (Germany), invertebrates were the most important scavengers of simulated kills in both summer and winter [13].

Eurasian lynx (*Lynx lynx*) in Central and Northern Europe, in contrast to their congeneric species and Eurasian lynx in Turkey, feeding mostly on smaller prey such as brown hares [24], have the ecology of far larger felids, mainly feeding on roe deer (*Capreolus capreolus*), which are 50–100% heavier than an adult lynx [6]. They normally feed on roe deer kills for periods of 2–7 days [8]. Vegetation and snow are used to cover kills during this period. We aimed to quantify the anti-scavenging effect of this covering behavior through three approaches. Firstly, we present observational data based on investigations of lynx-killed roe deer. Secondly, we present results from deer carcasses that were experimentally placed in the forest and video-monitored to determine the discovery time by vertebrate scavengers. Thirdly, we present the result of an experiment that recorded the rate at which invertebrate scavengers consumed carcass parts that were not accessible to vertebrate scavengers. In all cases, we compared discovery/consumption rates of carcasses that were covered with vegetation in the manner that lynx use, or left uncovered.

2. Materials and Methods

2.1. Study Area

The study was conducted in South-Eastern Norway, in Viken and Innlandet counties (formerly Hedmark, Østfold and Akershus counties). The northern part of the study area is characterized by extensive boreal forest with relatively low human population densities. The habitat in the southern part consists of intensively exploited boreal forest interspersed with agricultural land [25]. At the time of the study, Eurasian lynx were widespread in the landscape and mainly preyed on roe deer [6]. Important potential vertebrate scavengers were pine marten (*Martes martes*), badger (*Meles meles*), red fox (*Vulpes vulpes*), domestic cat (*Felis domesticus*), raven (*Corvus corax*), hooded crow (*Corvus corone cornix*), magpie (*Pica pica*) and Eurasian jay (*Garrulus glandarius*).

2.2. Lynx-Killed Roe Deer

The roe deer kills of very high frequency (VHF) radio-collared and snow-tracked lynx were routinely located in the period 1995–2008 [25,26]. Field procedures are described in Nilsen et al. 2009 [25]. Permissions were granted by the Norwegian Environment Agency and procedures were approved by the Norwegian Committee for Experimental Animal Welfare (permit numbers 08/127430, 07/81885, 07/7883, 2004/48647, 201/01/641.5/FHB, 127/03/641.5/fhb, 1460/99/641.5/FBe, 1081/97/641.5/FBe, and Norwegian Institute for Nature Research (NINA) 1/95). Most of the lynx-killed roe deer carcasses included in this study were found in winter (72 in winter and 7 in summer, Table S1). Kills were checked at various times. When studying predation behavior of collared lynx we only approached kills when the lynx had abandoned the kill. When snow-tracking unmarked lynx or trying to recapture already collared lynx, we would approach a new found kill straight away in order to place traps. When the kills were examined, any signs of the presence of scavengers were recorded (such as scats, footprints, feathers) as well as signs of the kill being covered by the lynx. However, we cannot exclude that some scavengers remained undetected. We knew lynx identity for 53 lynx-killed roe deer carcasses. For this subset of data, we explored individual variation in lynx caching behavior by comparing the proportion of covered carcasses for single individuals vs. females with kittens, and for males vs. females, by means of Chi-square tests. We had information on habitat type for 66 of the lynx-killed roe deer carcasses. For this subset of data, we compared the proportion of covered carcasses in open habitats and forest habitats, by means of Chi-square tests.

We used multivariate logistic regression models to explore which variables better explained the probability of a carcass being scavenged or not. The full model explaining the probability of a carcass being scavenged included number of days after being killed, the treatment (not covered/covered) and their interaction as explanatory variables. The final model was selected by model reduction, where only explanatory variables with significance $p < 0.05$ were retained in the model.

2.3. Experimentally-Deployed Roe Deer Carcasses

From 2002 to 2003, we monitored 26 experimentally-deployed whole roe deer carcasses (obtained as roadkill) as simulated lynx kills for 7 days using time-lapse video equipment (Table S2). The time-lapse video observations were made with a VHS tape recorder (Sanyo TLS 1960P) powered by two 12-volt batteries. This was connected to an infrared light (IR 70, 940 Nm), that was switched on 15 min before dusk and switched off 15 min after dawn, and a lens (Sony CWSHR WVF) with a miniTV cable (20 to 120 m long), so that the batteries and the tape could be changed every 24–72 h without disturbing the carcass site. We recorded 72 h on a 180 min videotape with a time-lapse of 2.5 frames per sec. The lens and IR light were mounted in a tree (2–3 m height) ca. 5 m from the carcass. The carcasses were deployed in forested habitat (in forest, small openings in forest, or forest edges). An equal number of kills were deployed in summer and winter seasons. Fourteen kills were completely covered with grass, moss or snow in the manner similar to lynx, and 12 were left completely exposed. For each kill, we determined the time of first arrival of mammalian and avian scavengers as a measure of time to detection. However, subsequent consumption rate by birds and mammals was not quantified, so detection time might not be directly related to consumption rate, although it is expected that scavengers detect the cues of other scavengers on a site [27]. Within seven days from deployment, four carcasses were not found by any scavengers, three were found only by mammals and five were found only by birds. Linear models were run to explore the factors influencing discovery time separately for birds and mammals, where the full model included treatment (covered/not covered), season (winter/summer) and their interaction as explanatory variables and number of days from deployment to discovery as dependent variable.

2.4. Invertebrate Scavenging Experiments

In summer 2003, we deployed two whole roe deer and eight roe deer body parts (obtained as roadkill animals, divided approximately into quarters for each treatment, with skin and hair in place). The five pairs of carcasses/carcass parts were placed in iron mesh cages (size $1 \times 1 \times 0.5$ m, mesh diameter 2.5 cm) in forest habitat to explore the rate of weight loss by invertebrate scavenging and decomposition while excluding all vertebrate scavengers above the size of a small rodent (Table S3). The experiment was designed to permit weighing the cage with only a minimal vertical lift to avoid losing tissue or dislodging parts of the carcass or the covering. A finer mesh (mesh diameter 1.9 cm) was also installed in the base of the cage, to minimize this issue, ensuring that only decomposition fluids and gasses were lost. However, it is important to be aware of the fact that the weight includes the eventual weight of insect eggs and maggots laid on, or feeding on, the carcass. Our scales were accurate to 100 g. We cannot exclude the possibility that small rodents might have removed some minor amounts of tissue. We used a paired experimental design, where one sample was covered with vegetation and another was left uncovered. Each sample was weighed daily (for a few missing days, we averaged the adjacent days) for 3 weeks following deployment. A four-parameter Weibull model, which represents an asymmetric sinusoidal decay [27], was fitted to the linear stretch of each subset of data by non-linear regression [28]. Three parameters were extracted to describe each curve: (1) day of start of decay (minimum value of the second order derivate of the curve); (2) day of inflection (minimum value of the first order derivate of the curve); (3) day of flattening out (maximum value of the second order derivate of the curve). The values for these three stages were compared using Mann–Whitney U tests.

Analyses were done in R [29], where the Weibull models were run with the add-on package drc [30].

3. Results

3.1. Lynx-Killed Roe Deer

We inspected 79 lynx-killed roe deer from one to eleven days after the kill was made. Eighteen of the 79 (23%) carcasses had tracks from one or several scavengers (thirteen, four, and one carcasses had signs of avian, mammalian and unknown scavengers, respectively). Lynx had covered 35 of 79 (44%) of the carcasses partially or totally with snow or a combination of live and dead plant material available at the kill-site. Only 9% of the covered carcasses vs. 28% of the uncovered carcasses were scavenged (Figure 1).

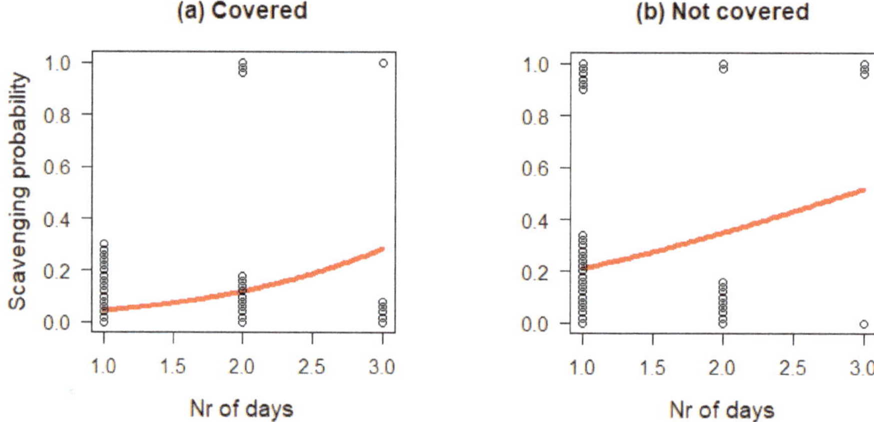

Figure 1. Logistic regression model curves explaining the probability of being scavenged in relation to the number of days since the kill was made for (**a**) covered and (**b**) not covered lynx-killed roe deer carcasses in South-Eastern Norway ($n = 74$).

Each identified lynx individual killed on average 2.21 ($\pm$1.35) of the 53 roe deer carcasses for which we had information on lynx identity. We could not detect differences in caching behavior between single individuals compared to family groups (female with kittens) (Chi-square = 1.93, df = 1, $p = 0.164$) or between males compared to females (Chi-square = 1.73, df = 1, $p = 0.188$). Similarly, we could not detect differences in caching behavior in open areas compared to forest (Chi-square = 0.433, df = 1, $p = 0.5105$).

The logistic model explaining the probability of being scavenged selected by model reduction included number of days after the kill and treatment as explanatory variables. The probability of being scavenged increased with number of days since the kill was made, and when the kill was uncovered. The regression equation for covered carcasses was y = 0.03 + 0.69 * Nr of days, whereas the regression equation for not covered carcasses was y = 0.80 + 0.69 * Nr of days ($p < 0.05$ for all terms).

3.2. Experimentally-Deployed Roe Deer Carcasses

For mammalian scavengers, the model reduction ended with a model including only the main effect of season with a positive effect of 'winter', implying a longer time to first discovery in winter (Adjusted $R^2 = 0.376$, $n = 26$, Table 1a; Figure 1). For avian scavengers, the model reduction included only the main effect of treatment with a negative effect of 'not covered', implying a longer time to first discovery of covered kills ($R^2 = 0.446$, $n = 26$, Table 1b, Figure 2).

Table 1. Estimates of model (selected by model reduction) explaining discovery time for (a) mammalian and (b) avian scavengers of 26 carcasses with season of placement, treatment (covered vs. not covered (NC)) and their interaction as explanatory variables.

	Estimate	SE	95% CI Lower	Upper	*p*-Value
(a) Mammals					
Summer	2.81	0.50	1.82	3.80	<0.001
Winter	3.13	0.71	1.73	4.53	<0.001
(b) Avian					
Covered	4.51	0.58	3.37	5.64	<0.001
Not covered	−3.25	0.85	−4.92	−1.57	<0.001

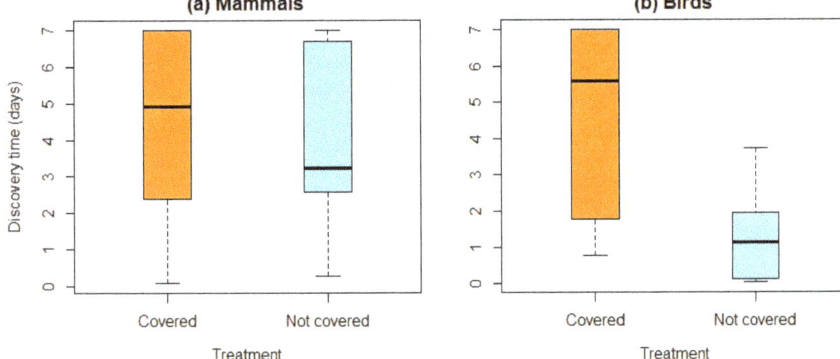

Figure 2. Box plots showing the effect of covering video-monitored experimentally-deployed roe deer carcasses on time to discovery (in days) for (**a**) mammalian, and (**b**) avian scavengers in South-Eastern Norway.

3.3. Insect Scavenging Experiments

Although in most cases the covered carcasses started to decay later than uncovered carcasses (*n* = 5 in each treatment), the difference between median dates of decay initiation of covered versus uncovered carcasses (6.27 vs. 4.27; ns) was not significant. Both the inflection point (8.41 vs. 5.41; *p* = 0.015) and day of flattening out of the decay curve (11.11 vs. 6.41; *p* = 0.0079) occurred significantly later for covered carcasses (Table 2, Figure 3), implying that covering slowed invertebrate consumption/decomposition.

Table 2. Values obtained by fitting a four-parameter Weibull model by non-linear regression to the consecutive weigh of the roe deer carcasses deployed to simulate lynx kills (NC: not covered; C: covered). In parentheses, it is reported the initial weight of the not covered/covered carcasses/carcass parts. Day of start of decay = minimum value of the second order derivate, inflection point = minimum value of the first order derivate and day of flattening out of the decay curve = maximum value of the second order derivate.

	Day of Start of Decay NC	C	Inflection Point NC	C	Day of Flattening Out NC	C
Whole carcass (25/23.2 kg)	5.47	18.05	7.57	20.01	9.59	21.89
August pair 1 (15/4.8 kg)	4.12	0.00	4.79	6.03	5.41	12.22
August pair 2 (3.2/3.3 kg)	4.26	4.07	5.38	6.84	6.48	9.45
September pair 1 (17.4/4 kg)	4.71	8.07	5.58	9.61	6.39	11.13
September pair 2 (3.3/3.2 kg)	4.04	6.28	5.02	8.41	5.94	10.46
Median	4.26	6.28	5.38	8.41	6.39	11.13

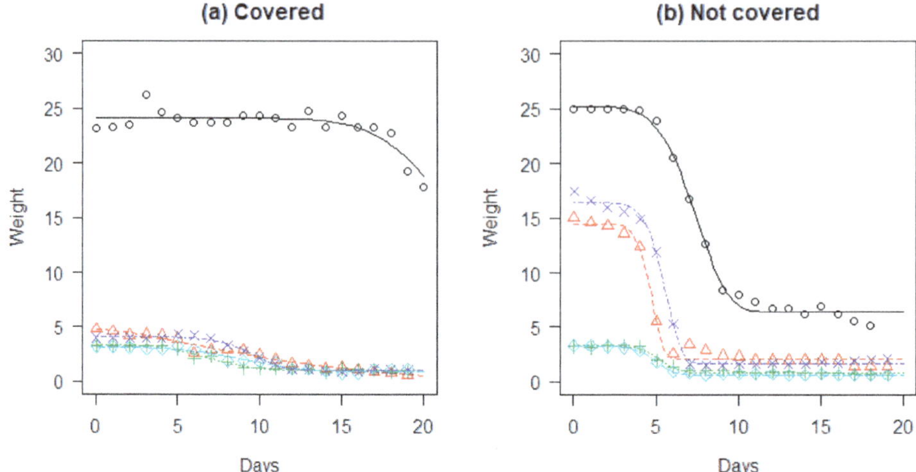

Figure 3. Daily weight variation (kg) and fitted Weibull curves describing the decay process of (**a**) five covered, and (**b**) five uncovered paired roe deer carcass or parts placed in a metal cage to exclude vertebrate scavengers during summer in South-Eastern Norway. The line in black in (**a**,**b**) are whole carcasses; the other lines are parts.

4. Discussion

This study provides clear evidence that the simple act of covering a kill with vegetation or snow delays scavenger arrival and the rate of meat loss. The lynx-killed roe deer that were uncovered were three times more likely to have been found by scavengers than the kills that were covered. Covering the video-monitored experimental carcasses delayed their discovery by avian scavengers by 4–5 days. Covering the carcass parts delayed the rate of consumption by invertebrates by more than five days, although this result should be taken with caution due to the limited sample size for the insect experiment. This implies that covering a carcass constitutes a very effective cache protection measure [13,31] for a large felid preying on ungulates in forested habitats. Our results are consistent with a similar study on mountain lion caching behavior in Arizona, which showed that simulated caching reduced wastage during dry and hot periods [21].

Although most of the lynx-killed roe deer carcasses were from the winter, and only ca. 10% were from the summer, it is likely that caching behavior is similar in the two seasons. In the summer, there is no snow to cover kills with, so lynx can only use vegetation, and there is more competition with invertebrate scavengers, making covering more important (see experiment). However, in the summer there are also more leaves on trees and bushes making visual detection by avian scavengers harder. In the summer, the olfactory cues due to insect and microorganism activity might also increase the probability of a carcass being discovered. These aspects are not likely to affect the overall relative utility of covering kills vs. leaving them uncovered.

Although other authors have described this tendency of large felids to cover their cached kills with vegetation e.g., [5,13,21], this is the first study to actually test the anti-scavenger efficacy of this measure in the field, using both observational data from documented lynx kills and experimental approaches. Combined, these studies are revealing insights into the subtle ways that predator behavior responds to the potential for food loss from both vertebrate and invertebrate scavengers [11].

The results from the video-monitoring suggest that covering was efficient for deterring avian, but not mammal, scavengers. This indicates that the effect is mainly through visual occlusion, rather than by reducing the olfactory signal of the kills. The effect on invertebrate activity is probably through a process of physically obstructing invertebrate access to the kill, although the covering may also shade the kill and reduce decomposition speed by reducing the temperature. In a similar study conducted in the Bavarian forest in summer, roe deer carcasses simulating lynx kills (covered by vegetation) were completely consumed within 10 days, mostly by invertebrates [13].

Although the main focus of our study was to test the efficacy of caching, it is interesting to note that considerable periods of time, measured in days, went by before scavengers detected the kills. In contrast, studies on cheetahs in savannah ecosystems have shown that scavengers find kills within hours of death [20]. The effect is that many cheetah kills are either lost to kleptoparasites or are so rapidly consumed by scavengers that cheetahs are not able to consume more than one meal per kill. The lynxes in our study area, by comparison, were usually able to completely consume most of the meat on their kills over several days. This allows them to continue being roe deer specialists even when roe deer occur at low density [31], because even though there may be considerable search time expended in finding a prey, once killed it can provide food for many days. In turn, this can help explain why lynx presence has such a clear impact on low-density roe deer populations [32,33]. This efficient use of prey also implies that the energy budget of lynx should be considerably buffered against changes in prey density. The results also provide insight into the ways in which studying predator–prey dynamics needs to consider wider ecosystem processes.

At the time of our study, the area lacked the large scavengers such as wolverine (*Gulo gulo*), brown bear (*Ursus arctos*), wolf (*Canis lupus*) and wild boar (*Sus scrofa*), which could potentially drive a lynx off its kill and appropriate the kill in a single act. However, a study conducted in Sweden found no evidence of kleptoparasitism by wolves on roe deer killed by lynx [34]. The largest scavengers we documented were humans who regularly removed lynx kills [35], not an uncommon practice on a global scale [36]. It is hard to know how the absence of these natural members of the mammalian scavenger guild influences the generality of our results. A study conducted in Slovenia and Croatia found that brown bears were able to discover 32% of lynx prey remains, and 15% of all biomass of large prey killed by lynx was lost to bears [37]. Since this study was conducted, wolves and wild boar have colonized the landscape, opening for future comparative studies. However, the situation does reflect the current reality for lynx throughout much of its present day distribution in Europe [38,39], and the literature does tend to place greatest emphasis on the role of avian scavengers [16,40] for which our results indicate that the anti-scavenging behavior had the greatest effect.

Supplementary Materials: The following are available online at http://www.mdpi.com/1424-2818/12/9/350/s1, Table S1: Data from 79 lynx-killed roe deer inspected after the kill was made (South-Eastern Norway, 1995–2008). Table S2: Data from 26 experimentally-deployed whole roe deer carcasses (obtained as roadkill, South-Eastern Norway 2002–2003), used as simulated lynx kills for seven days using time-lapse video equipment. Table S3: Data from two whole roe deer and eight roe deer body parts (placed in iron mesh cages (size $1 \times 1 \times 0.5$ m, mesh diameter 2.5 cm) in forest habitat during summer (South-Eastern Norway, 2003) to explore the rate of weight loss by invertebrate scavenging and decomposition.

Author Contributions: Conceptualization, J.D.C.L. and I.J.M.T.; methodology, J.D.C.L., C.M., J.O. and I.J.M.T.; formal analysis, C.M., J.O. and I.J.M.T.; writing—original draft preparation, J.D.C.L., C.M., J.O. and I.J.M.T.; writing—review and editing, C.M., J.D.C.L., J.O. and I.J.M.T.; funding acquisition, J.D.C.L., J.O. and I.J.M.T. All authors have read and agreed to the published version of the manuscript.

Funding: This study is part of the SCANDLYNX project, and is funded by the Norwegian Environment Agency, the Research Council of Norway (two individual travel scholarships to IJM Teurlings and projects 134242, 165814, 183176, 212919, 251112), the county governor's office of Viken county, the Norwegian University for Science and Technology, and the Schure-Beijerinck-Popping Fund.

Acknowledgments: We would like to thank Ivar Herfindal for his statistical advice, Alan Bryan, Jan van Mil and all the other people who helped with the fieldwork.

Conflicts of Interest: The authors declare no conflict of interest.

References

1. Solomon, M.E. The Natural Control of Animal Populations. *J. Anim. Ecol.* **1949**, *18*, 1–35. [CrossRef]
2. Gervasi, V.; Nilsen, E.B.; Sand, H.; Panzacchi, M.; Rauset, G.R.; Pedersen, H.C.; Kindberg, J.; Wabakken, P.; Zimmermann, B.; Odden, J.; et al. Predicting the potential demographic impact of predators on their prey: A comparative analysis of two carnivore-ungulate systems in Scandinavia. *J. Anim. Ecol.* **2012**, *81*, 443–454. [CrossRef] [PubMed]
3. Vucetich, J.A.; Peterson, R.O.; Schaefer, C.L. The effect of prey and predator densities on wolf predation. *Ecology* **2002**, *83*, 3003–3013. [CrossRef]
4. Hayward, M.W.; Henschel, P.; O'Brien, J.; Hofmeyr, M.; Balme, G.; Kerley, G.I.H. Prey preferences of the leopard (Panthera pardus). *J. Zool.* **2006**, *270*, 298–313. [CrossRef]
5. Laundre, J.W.; Hernandez, L. Winter hunting habitat of pumas *Puma concolor* in northwestern Utah and southern Idaho, USA. *Wildl. Biol.* **2003**, *9*, 123–129. [CrossRef]
6. Odden, J.; Linnell, J.D.C.; Andersen, R. Diet of Eurasian lynx, *Lynx lynx*, in the boreal forest of southeastern Norway: The relative importance of livestock and hares at low roe deer density. *Eur. J. Wildl. Res.* **2006**, *52*, 237–244. [CrossRef]
7. Melovski, D.; Ivanov, G.; Stojanov, A.; Avukatov, V.; Gonev, A.; Pavlov, A.; Breitenmoser, U.; von Arx, M.; Filla, M.; Krofel, M.; et al. First insight into the spatial and foraging ecology of the critically endangered Balkan lynx (*Lynx lynx balcanicus*, Buresh 1941). *Hystrix Ital. J. Mammal.* **2020**, *31*, 26–34. [CrossRef]
8. Schmidt, K. Variation in daily activity of the free-living Eurasian lynx (*Lynx lynx*) in Bialowieza Primeval Forest, Poland. *J. Zool.* **1999**, *249*, 417–425. [CrossRef]
9. Sunquist, M.E.; Sunquist, F.C. Ecological constraints on predation by large felids. In *Carnivore Behavior, Ecology, and Evolution*; Gittleman, J.L., Ed.; Cornell University Press: New York, NY, USA, 1989; pp. 283–301.
10. DeVault, T.L.; Rhodes, O.E.; Shivik, J.A. Scavenging by vertebrates: Behavioral, ecological, and evolutionary perspectives on an important energy transfer pathway in terrestrial ecosystems. *Oikos* **2003**, *102*, 225–234. [CrossRef]
11. Mattisson, J.; Andren, H.; Persson, J.; Segerstrom, P. Influence of intraguild interactions on resource use by wolverines and Eurasian lynx. *J. Mammal.* **2011**, *92*, 1321–1330. [CrossRef]
12. Melis, C.; Teurlings, I.; Linnell, J.D.C.; Andersen, R.; Bordoni, A. Influence of a deer carcass on Coleopteran diversity in a Scandinavian boreal forest: A preliminary study. *Eur. J. Wildl. Res.* **2004**, *50*, 146–149. [CrossRef]
13. Ray, R.R.; Seibold, H.; Heurich, M. Invertebrates outcompete vertebrate facultative scavengers in simulated lynx kills in the Bavarian Forest National Park, Germany. *Anim. Biodivers. Conserv.* **2014**, *37*, 77–88.
14. Selva, N.; Fortuna, M.A. The nested structure of a scavenger community. *Proc. Roy. Soc. B Biol. Sci.* **2007**, *274*, 1101–1108. [CrossRef] [PubMed]
15. Selva, N.; Jedrzejewska, B.; Jedrzejewski, W.; Wajrak, A. Scavenging on European bison carcasses in Bialowieza Primeval Forest (eastern Poland). *Ecoscience* **2003**, *10*, 303–311. [CrossRef]
16. Kaczensky, P.; Hayes, R.D.; Promberger, C. Effect of raven *Corvus corax* scavenging on the kill rates of wolf *Canis lupus* packs. *Wildl. Biol.* **2005**, *11*, 101–108. [CrossRef]
17. Gorman, M.L.; Mills, M.G.; Raath, J.P.; Speakman, J.R. High hunting costs make African wild dogs vulnerable to kleptoparasitism by hyaenas. *Nature* **1998**, *391*, 479–481. [CrossRef]
18. Van der Veen, B.; Mattisson, J.; Zimmermann, B.; Odden, J.; Persson, J. Refrigeration or anti-theft? Food-caching behavior of wolverines (*Gulo gulo*) in Scandinavia. *Behav. Ecol. Sociobiol.* **2020**, *74*. [CrossRef]
19. Vucetich, J.A.; Peterson, R.O.; Waite, T.A. Raven scavenging favours group foraging in wolves. *Anim. Behav.* **2004**, *67*, 1117–1126. [CrossRef]
20. Hunter, J.S.; Durant, S.M.; Caro, T.M. Patterns of scavenger arrival at cheetah kills in Serengeti National Park Tanzania. *Afr. J. Ecol.* **2007**, *45*, 275–281. [CrossRef]
21. Bischoff-Mattson, Z.; Mattson, D. Effects of Simulated Mountain Lion Caching on Decompsition of Ungulate Carcasses. *West. N. Am. Nat.* **2009**, *69*, 343–350. [CrossRef]
22. Jobin, A.; Molinari, P.; Breitenmoser, U. Prey spectrum, prey preference and consumption rates of Eurasian lynx in the Swiss Jura Mountains. *Acta Theriol.* **2000**, *45*, 243–252. [CrossRef]
23. Vander Wall, S.B. *Food Hoarding in Animals*; University of Chicago Press: Chicago, IL, USA, 1990.

24. Mengulluoglu, D.; Ambarli, H.; Berger, A.; Hofer, H. Foraging ecology of Eurasian lynx populations in southwest Asia: Conservation implications for a diet specialist. *Ecol. Evol.* **2018**, *8*, 9451–9463. [CrossRef] [PubMed]

25. Nilsen, E.B.; Linnell, J.D.C.; Odden, J.; Andersen, R. Climate, season, and social status modulate the functional response of an efficient stalking predator: The Eurasian lynx. *J. Anim. Ecol.* **2009**, *78*, 741–751. [CrossRef] [PubMed]

26. Bunnefeld, N.; Linnell, J.D.C.; Odden, J.; van Duijn, M.A.J.; Andersen, R. Risk taking by Eurasian lynx (*Lynx lynx*) in a human-dominated landscape: Effects of sex and reproductive status. *J. Zool.* **2006**, *270*, 31–39. [CrossRef]

27. France, J.; Dijkstra, J.; Dhanoa, M.S. Growth functions and their application in animal science. *Anim. Res.* **1996**, *45*, 165–174. [CrossRef]

28. Venables, W.N.; Ripley, B.D. *Modern Applied Statistics with S*, 4th ed.; Springer: New York, NY, USA, 2002.

29. Team, R.C. *R: A Language and Environment for Statistical Computing*; R Foundation for Statistical Computing: Vienna, Austria, 2013.

30. Ritz, C.; Streibig, J.C. Bioassay analysis using R. *J. Stat. Softw.* **2005**, *12*, 1–22. [CrossRef]

31. Dally, J.M.; Clayton, N.S.; Emery, N.J. The behaviour and evolution of cache protection and pilferage. *Anim. Behav.* **2006**, *72*, 13–23. [CrossRef]

32. Melis, C.; Basille, M.; Herfindal, I.; Linnell, J.D.C.; Odden, J.; Gaillard, J.M.; Hogda, K.A.; Andersen, R. Roe deer population growth and lynx predation along a gradient of environmental productivity and climate in Norway. *Ecoscience* **2010**, *17*, 166–174. [CrossRef]

33. Melis, C.; Jedrzejewska, B.; Apollonio, M.; Barton, K.A.; Jedrzejewski, W.; Linnell, J.D.C.; Kojola, I.; Kusak, J.; Adamic, M.; Ciuti, S.; et al. Predation has a greater impact in less productive environments: Variation in roe deer, *Capreolus capreolus*, population density across Europe. *Glob. Ecol. Biogeogr.* **2009**, *18*, 724–734. [CrossRef]

34. Wikenros, C.; Liberg, O.; Sand, H.; Andren, H. Competition between recolonizing wolves and resident lynx in Sweden. *Can. J. Zool.* **2010**, *88*, 271–279. [CrossRef]

35. Krofel, M.; Kos, I.; Linnell, J.D.C.; Odden, J.; Teurlings, I.J.M. Human kleptoparasitism on Eurasian lynx (*Lynx lynx* L.) in Slovenia and Norway. *Varsto Narave* **2008**, *21*, 93–103.

36. Treves, A.; Naughton-Treves, L. Risk and opportunity for humans coexisting with large carnivores. *J. Hum. Evol.* **1999**, *36*, 275–282. [CrossRef] [PubMed]

37. Krofel, M.; Kos, I.; Jerina, K. The noble cats and the big bad scavengers: Effects of dominant scavengers on solitary predators. *Behav. Ecol. Sociobiol.* **2012**, *66*, 1297–1304. [CrossRef]

38. Linnell, J.D.C.; Breitenmoser, U.; Breitenmoser-Würsten, C.; Odden, J.; von Arx, M. Recovery of Eurasian lynx in Europe: What part has reintroduction played? In *Reintroduction of Top-Order Predators*; Hayward, M.W., Somers, M.J., Eds.; Wiley-Blackwell: Oxford, UK, 2009; pp. 72–91.

39. Linnell, J.D.C.; Promberger, C.; Boitani, L.; Swenson, J.E.; Breitenmoser, U.; Andersen, R. The linkage between conservation strategies for large carnivores and biodiversity: The view from the "half-full" forests of Europe. In *Carnivorous Animals and Biodiversity: Does Conserving One Save the Other?* Ray, J.C., Redford, K.H., Berger, J., Steneck, R.S., Eds.; Island Press: Washington, DC, USA, 2005; pp. 381–398.

40. Selva, N.; Jedrzejewska, B.; Jedrzejewski, W.; Wajrak, A. Factors affecting carcass use by a guild of scavengers in European temperate woodland. *Can. J. Zool.* **2005**, *83*, 1590–1601. [CrossRef]

 *diversity*

Article

Lack of Cascading Effects of Eurasian Lynx Predation on Roe Deer to Soil and Plant Nutrients

Ivonne J. M. Teurlings [1,†], Claudia Melis [2], Christina Skarpe [1,3] and John D. C. Linnell [1,*]

1   Norwegian Institute for Nature Research, PO Box 5685 Torgarden, NO-7485 Trondheim, Norway; i_teurlings@hotmail.com (I.J.M.T.); christina.skarpe@inn.no (C.S.)
2   Department of Nature, Environment and Health, Queen Maud University College for Early Childhood Education, Thrond Nergaards veg 7, 7044 Trondheim, Norway; cme@dmmh.no
3   Faculty of Applied Ecology, Agricultural Sciences and Biotechnology, Inland Norway University of Applied Sciences, Postboks 400, 2418 Elverum, Norway
*   Correspondence: john.linnell@nina.no
†   Present address: Saghaugen gård, Skjomdalveien-Bogholm 21, 8523 Skjomen, Norway.

Received: 13 August 2020; Accepted: 11 September 2020; Published: 14 September 2020

Abstract: This study examines the extent to which above-ground trophic processes such as large carnivore predation on wild ungulates can cause cascading effects through the provision of carrion resources to below-ground ecosystem processes in the boreal forest of southeastern Norway. We measured the levels of 10 parameters in soil samples and 7 parameters in vegetation (wavy hair-grass, *Avenella flexuosa*, and bilberry, *Vaccinium myrtillus*) at 0, 0.5 and 2 m distance from 18 roe deer (*Capreolus caprelous*) carcasses killed by Eurasian lynx (*Lynx lynx*). We then compared these values to two control sites 20 m away from each carcass. Sampling was conducted 20–29 months after death. Neither soil nor vegetation samples showed a clear gradient in parameters (CN, NH_4^+, NO_3^-, P, PO_4^-, Ca, K, Mg and Na) from the center of a carcass towards the periphery. Similarly, there was no difference in the effect on soil and vegetation between winter- and summer-killed carcasses. Our results contrast with that of other studies that simulate the effect of predation with whole carcasses and which often exclude scavengers through fencing. The lack of detectable effects after about two years is likely due to the small size of roe deer carcasses and the fact that most tissues are consumed by the predator and scavengers before decomposition.

Keywords: *Capreolus capreolus*; carrion; decomposition; *Lynx lynx*; nitrogen; nutrient recycling; trophic cascade

---

1. Introduction

There has been a focus on the wider ecological effects of large terrestrial carnivores in recent years driven by both the study of ecosystems where they are being reintroduced or are naturally increasing in density [1,2] and where their absence is conspicuous [3]. This focus is largely due to their putative role as ecological keystone species [4] or as strongly interactive species [5]. Although these impacts are not as universal as is often claimed [4,6], there are plenty of examples of large carnivores having cascading effects on the ecosystems of which they are a part [7,8]. These impacts mainly operate through the influence of large terrestrial carnivores on the density, distribution and behavior of their wild ungulate prey, species that have effects on vegetation through diverse mechanisms [3,9,10]. This in turn affects a wide range of other species that respond to the environmental changes [3,11]. Other mechanisms involve the increased availability of carrion to scavengers as a by-product of predation [12–17].

Another pathway through which predation can potentially influence diverse ecosystem functions is through the transfer of nutrients from carnivore-killed prey to soil and then into vegetation [9]. Some studies have shown that this nutrient pulse, although local in extent, can be intense [18–22].

For example, a range of studies have focused on the role of bears, *Ursus* spp., in transferring marine nutrients from anadromous salmon into terrestrial ecosystems [23]. However, the generality of this impact is likely to depend on how much the predator consumes its prey and how much of the remains are consumed or dispersed by scavengers (both micro and macro [24]) as it is only the leachate fluids resulting from invertebrate maggot activity and microbial decomposition that are directly transferred to soil [25]. In addition, nutrients also can originate from dead invertebrates, and excretion from both scavengers and predators feeding on or attracted to the carcass [25,26]. By this mechanism, the impacts of large carnivore predation can cascade through the ecosystem, bridging the divide between above- and below-ground ecosystems [27]. Moreover, carcasses, by locally reducing herbaceous cover, can influence the early stages of tree reproduction [28]. The impacts of carrion on soil and vegetation parameters are also influenced by temperature and the season of death [29–31].

The aim of our study was to investigate the presence of these cascading effects in a previously unstudied system. For this purpose, we studied a large carnivore (Eurasian lynx, *Lynx lynx*)–ungulate (roe deer, *Capreolus capreolus*) system in the boreal forest of southeastern Norway.

Our predictions were as follows:

(1) Lynx-killed roe deer carcasses will have an effect on nutrient levels in the surrounding soil and vegetation, so that soil and vegetation parameters would be changed to a greater extent closer to the center of the carcass and decline farther away from the carcass [21].

(2) This effect is different for kills made in winter and summer, where the parameters are more elevated for summer lynx-killed roe deer, because the absence of snow cover and freezing temperatures in summer permits a more rapid and direct downwards nutrient transfer [29–31].

2. Materials and Methods

2.1. Study Area

The study was conducted approximately 50 km southeast of Oslo, Norway, (N 59°36′55″, E 11°36′45″) in the municipalities of Spydeberg, Enebakk, Hobøl and Aurskog-Høland in Viken county (formerly Østfold and Akershus counties). The area consists of rolling hills with residential and agricultural areas (25%) surrounded by forest (75%) dominated by Norway spruce (*Picea abies*) and Scots pine (*Pinus sylvestris*) mixed with downy birch (*Betula pubescens*), mainly used for timber production. During the growing season, the agricultural fields are used for the production of cereals and fodder grass. The main ungulate species were roe deer and moose (*Alces alces*). The main predators of ungulates were lynx, which were present throughout the landscape, and individual transient wolves (*Canis lupus*). Important scavengers were pine martens (*Martes martes*), red foxes (*Vulpes vulpes*), Eurasian badgers (*Meles meles*), domestic cats (*Felis domesticus*), ravens (*Corvus corax*), hooded crows (*Corvus cornix*), magpies (*Pica pica*) and Eurasian jays (*Garrulus glandarius*), as found from video observation (unpublished data).

2.2. Study Design

The ecology of lynx predation on roe deer has been studied in the region since 2000 by (1) following VHF- and GPS-collared lynx to find kills, (2) snow-tracking unmarked lynx to find kills and (3) regularly monitoring marked roe deer to determine the timing and cause of death [32,33]. Field procedures are described in Nilsen et al. 2009 [32]. This work led us to locate and mark the kill-sites of roe deer that had been killed and at least partially consumed by lynx. Permissions were granted by the Norwegian Environment Agency and procedures were approved by the Norwegian Committee for Experimental Animal Welfare (permit numbers 08/127430, 07/81885, 07/7883, 2004/48647, 201/01/641.5/FHB, 127/03/641.5/fhb, 1460/99/641.5/FBe, 1081/97/641.5/FBe, and NINA 1/95).

Soil and vegetation samples were simultaneously collected at the end of the growing season (between the 29th of June and the 29th of July) in 2004 from 18 carcass sites, where six of the roe deer were killed by lynx in winter (October–April) and 12 in summer (May–September). The samples were taken 20–29 months after death, following the methods described in Melis et al. [20]. The samples were taken along a line with the carcass as the center, at distances of 0, 0.5, 2 and 20 m extending in opposite directions from the center of the carcass. The head to tail length of a roe deer is between 95 and 135 cm, and lynx and other scavengers can move roe deer carcasses slightly while feeding on them, so the distances from the center to 0.5 m were considered as being directly influenced by the carcass. The sample at 20 m was considered not to be affected by the carcass and was used as a control. When we visited the carcass site after the lynx had left (Figure 1a,b), we documented how much of the kill was left, as well as ground vegetation and forest type. We marked the carcass site with a pole and a label (Figure 1a,c,d) at the middle point and took pictures to be able to find again the exact carcass position. In 16 of the 18 carcasses, the lynx had removed 90% of the meat according to our visual estimation several days after the lynx had abandoned the kill. The direction of the sample line was chosen so that it fell parallel to the contour of the landscape, rather than perpendicular to the slope. The samples from both sides were merged for each distance for analysis. The control sample at 20 m was taken within the same vegetation type as the carcass was at. The litter was removed and a 2 × 2 × 5 cm (length × width × depth) sample of soil was taken from the surface. Fresh green leaves were collected (3–5 g dry weight per sample) from the most common plants present at the sites (wavy hair-grass, *Avenella flexuosa*, and bilberry, *Vaccinium myrtillus*). The combination of nutrient leaching, pH shock (a rapid increase in pH) and blocking of light and oxygen by the physical presence of the carcass killed the vegetation directly under the carcass (Figure 1c), as described in [25]. This effect could last for more than two years (Figure 1d). Therefore, we were not always able to collect vegetation samples at the center.

Figure 1. Roe deer carcass sites in southeastern Norway. (**a**,**b**) Carcass sites visited as soon as the lynx left the kill. Both the carcasses had been fed on by lynx for four nights, and the images were taken two and six days after the lynx had abandoned the kill. (**c**,**d**) The same carcass site (**c**) one month and (**d**) two years ca. after death. In (**a**,**c**,**d**), it is possible to see the labeled pole marking the center of the carcass.

2.3. Laboratory Analyses

In total, 70 soil and 52 vegetation (25 *A. flexuosa*, 27 *V. myrtillus*) samples were analyzed by the Wageningen University Testing Laboratory of the Resource Ecology Group.

In soil, the concentrations of total nitrogen (N), total phosphorus (P) and calcium (Ca) were determined after digestion with H_2SO_4 ⁻Se-salicylacid-H_2O_2 [34,35]. Concentrations of inorganic nitrogen (nitrate NO_3^-, ammonium NH_4^+), inorganic phosphorus (phosphate PO_4^-), potassium (K), magnesium (Mg) and sodium (Na) in soil were determined after extraction with 0.01 M $CaCl_2$ by using the method described in [36].

In plant leaves, the concentrations of total N, total P, Ca, K, Mg and Na were determined by digestion with H_2SO_4 ⁻Se-salicylacid-H_2O_2. The C/N ratio in both soil and vegetation was determined using a C/N analyzer (Fisons EA 1108, CHN-0, Woodstock III, Interscience B.V., Breda, Brabant, The Netherlands). Concentrations of total N and total P in soil and vegetation, and inorganic nitrogen (nitrate NO_3^-, ammonium NH_4^+) and inorganic phosphorus (phosphate PO_4^-) in soil were measured with an auto-analyzer (Skalar San Plus, colorimetric system, Skalar Analytical B.V., Breda, Noord-Brabant, The Netherlands). K, Ca, Mg and Na were measured with an atomic absorption spectrometer (Varian, SpectrAA-600, Varian B.V. Benelux, Middelburg, Zeeland, The Netherlands).

2.4. Statistical Analyses

To account for statistical non-independence caused by having several samples at the same carcass site, the variable "site" was added as a random effect in a generalized linear mixed model (GLMM) framework [37]. We used a model selection procedure based on the Akaike information criterion corrected for small sample sizes [38] to identify which variables were best at explaining the variation in the concentration of a particular nutrient. The full model included the concentration of a nutrient as the dependent variable, and "distance" (distance from the center of the carcass) and "season" (season of kill) as explanatory variables. The minimum adequate model (MAM, hereafter) for each analysis was chosen based on the AICc value, Akaike weight and the number of parameters [38]. If the variable "distance" was not included in the MAM, we assumed that the effect of the carcass was not detectable on that parameter.

Shapiro–Wilk normality tests and Q–Q plots were used to check for normality. Response variables were transformed to improve normality, when not normally distributed. For soil nutrient concentrations, we square root-transformed nitrogen, calcium and magnesium, cube root-transformed phosphorus and sodium and log-transformed (natural logarithm) all other parameters. For vegetation samples, we square root-transformed calcium and magnesium, cube root-transformed sodium and log-transformed all other parameters.

The analyses were conducted separately for soil and for each of the two plant species using R 2.6.1 software [39], where the model selection was done with the R package MuMIn version 1.43.17 [40] and the linear mixed models were run with the add-on package lme4 [41]. The mixed models were fitted with maximum likelihood for the calculation of the AICc values, while the parameter estimates and their standard errors were from models fitted with restricted maximum likelihood [37].

3. Results

The variable "distance" was not included in the MAM for either soil or vegetation nutrients (data on model selection shown in Table 1 for soil, and in Tables 2 and 3 for *V. myrtillus*, and *A. flexuosa*, respectively). This indicated that the effect of the carcasses on soil and vegetation nutrient concentrations could not be detected (Figures 2–4).

Table 1. Set of linear mixed effect models with concentration (g/100 g) of minerals in soil in southeastern Norway in 2004 as the dependent variable and winter vs. summer (Season) and distance from the center of the carcass (Distance) as the explanatory variables. (a) C/N ratio; (b) nitrogen, N; (c) ammonium, NH_4+; (d) nitrate, NO_3-; (e) phosphorus, P; (f) phosphate, PO_4-; (g) calcium, Ca; (h) potassium, K; (i) magnesium, Mg; (j) sodium, Na. The plus sign (+) indicates that the factor was included in the model.

	Intercept	Season	Distance [a]	K [b]	AICc	Δ AICc [c]	ωi [d]
(a) C/N ratio (log)	2.95			3	−38.70	0.00	0.634
	2.87	+		4	−37.50	1.15	0.357
	2.94		0.01	4	−29.20	9.50	0.006
	2.86	+	0.01	5	−28.00	10.70	0.003
(b) N (square root)	0.92			3	−35.40	0.00	0.895
	0.92	+		4	−30.60	4.80	0.081
	0.94		−0.01	4	−27.90	7.47	0.021
	0.95	+	−0.01	5	−23.00	12.32	0.002
(c) NH_4+ (log)	−6.25			3	120.20	0.00	0.670
	−6.14	+		4	121.90	1.74	0.281
	−6.18		−0.04	4	126.10	5.94	0.034
	−6.08	+	−0.04	5	127.90	7.73	0.014
(d) NO_3^- (log)	−9.18			3	231.90	0.00	0.565
	−9.11	+		4	232.90	1.01	0.341
	−9.05		−0.08	4	236.40	4.50	0.060
	−8.98	+	−0.08	5	237.40	5.58	0.035
(e) P (cube root)	0.41			3	−241.60	0.00	0.884
	0.40	+		4	−237.50	4.13	0.112
	0.42		0.00	4	−230.30	11.29	0.003
	0.40	+	0.00	5	−226.00	15.56	0.000
(f) PO_4^- (log)	−7.60	+		4	261.90	0.00	0.491
	−7.91			3	262.40	0.44	0.394
	−7.44	+	−0.09	5	266.00	4.11	0.063
	−7.75		−0.09	4	266.40	4.49	0.052
(g) Ca (square root)	0.69			3	−130.50	0.00	0.945
	0.69	+		4	−124.40	6.02	0.047
	0.70		−0.01	4	−120.90	9.59	0.008
	0.70	+	−0.01	5	−114.80	15.69	0.000
(h) K (log)	−4.25			3	145.20	0.00	0.661
	−4.14	+		4	146.80	1.57	0.301
	−4.20		−0.03	4	151.70	6.44	0.026
	−4.09	+	−0.03	5	153.30	8.07	0.012
(i) Mg (square root)	0.13			3	−285.60	0.00	0.983
	0.13	+		4	−276.60	9.01	0.011
	0.13		0.00	4	−275.30	10.27	0.006
	0.14	+	0.00	5	−266.30	19.34	0.000
(j) Na (cube root)	0.19			3	−160.00	0.00	0.865
	0.21	+		4	−156.10	3.90	0.123
	0.20		−0.01	4	−151.00	8.95	0.010
	0.22	+	−0.01	5	−147.10	12.89	0.001

[a] The explanatory variable "Distance" was square root-transformed. The models were ranked by the AICc—corrected Akaike information criterion (AIC). The minimum adequate model is on the top of each list; [b] K, number of parameters; [c] Δ AICc, difference in Akaike values between the first and the actual model; [d] ωi, Akaike weights.

Table 2. Set of linear mixed effect models with concentration of minerals (g/100 g) in *Vaccinium myrtillus* in southeastern Norway in 2004 as the dependent variable and winter vs. summer (Season) and distance from the center of the carcass (Distance) as the explanatory variables. (a) C/N ratio; (b) nitrogen, N; (c) phosphorus, P; (d) potassium, K; (e) calcium, Ca; (f) magnesium, Mg; (g) sodium, Na. The plus sign (+) indicates that the factor was included in the model. For further abbreviations, see Table 1.

	Intercept	Season	Distance	K	AICc	Δ AICc	ωi
(a) C:N ratio (log)	3.37			3	−5.00	0.000	0.818
	3.33	+		4	−1.90	3.140	0.170
	3.34		0.01	4	3.70	8.720	0.010
	3.29	+	0.01	5	7.00	12.090	0.002
(b) N (square root)	1.29			3	−8.50	0.000	0.896
	1.27	+		4	−3.70	4.770	0.082
	1.35		−0.02	4	−0.90	7.640	0.020
	1.33	+	−0.02	5	4.20	12.720	0.002
(c) P (cube root)	0.49			3	−90.80	0.000	0.939
	0.51		−0.01	4	−84.90	5.940	0.048
	0.49	+		4	−82.20	8.650	0.012
	0.51	+	−0.01	5	−75.80	15.000	0.001
(d) Ca (square root)	0.91			3	−59.00	0.000	0.958
	0.90	+		4	−52.40	6.590	0.035
	0.93		−0.01	4	−49.10	9.890	0.007
	0.92	+	−0.01	5	−42.20	16.790	0.000
(e) K (log)	−0.26			3	29.50	0.000	0.868
	−0.28	+		4	33.60	4.130	0.110
	−0.28		0.01	4	37.00	7.570	0.020
	−0.30	+	0.01	5	41.40	11.950	0.002
(f) Mg (square root)	0.46			3	−87.30	0.000	0.968
	0.45	+		4	−80.20	7.050	0.029
	0.47		0.00	4	−75.90	11.370	0.003
	0.46	+	0.00	5	−68.50	18.820	0.000
(g) Na (cube root)	0.13			3	−27.90	0.000	0.931
	0.15	+		4	−22.40	5.470	0.060
	0.15		−0.01	4	−18.40	9.490	0.008
	0.17	+	−0.01	5	−12.70	15.200	0.000

Table 3. Set of linear mixed effect models with concentration of minerals (g/100 g) in *Avenella flexuosa* in southeastern Norway in 2004 as the dependent variable and winter vs. summer (Season) and distance from the center of the carcass (Distance) as the explanatory variables. (a) C/N ratio; (b) nitrogen, N; (c) phosphorus, P; (d) potassium, K; (e) calcium, Ca; (f) magnesium, Mg; (g) sodium, Na. The plus sign (+) indicates that the factor was included in the model. For further abbreviations, see Table 1.

	Intercept	Season	Distance	K	AICc	Δ AICc	ωi
(a) C/N ratio (log)	3.42			3	6.30	0.000	0.903
	3.40	+		4	11.30	4.940	0.076
	3.36		0.02	4	14.10	7.730	0.019
	3.34	+	0.02	5	19.40	13.060	0.001
(b) N (square root)	1.20			3	−4.50	0.000	0.935
	1.20	+		4	1.30	5.800	0.051
	1.17		0.01	4	4.00	8.530	0.013
	1.16	+	0.01	5	10.10	14.570	0.001
(c) P (cube root)	0.50			3	−59.20	0.000	0.938
	0.49	+		4	−53.60	5.570	0.058
	0.49		0.00	4	−48.20	10.920	0.004
	0.48	+	0.00	5	−42.20	16.970	0.000
(d) Ca (square root)	0.36			3	−53.80	0.000	0.948
	0.34	+		4	−47.80	5.960	0.048
	0.36		0.00	4	−42.70	11.080	0.004
	0.35	+	0.00	5	−36.50	17.240	0.000
(e) K (log)	0.73			3	16.20	0.000	0.894
	0.74	+		4	20.80	4.620	0.089
	0.69		0.01	4	24.30	8.100	0.016
	0.71	+	0.01	5	29.10	12.960	0.001
(f) Mg (square root)	0.30			3	−56.80	0.000	0.976
	0.30	+		4	−48.80	7.970	0.018
	0.32		−0.01	4	−46.50	10.280	0.006
	0.32	+	−0.01	5	−38.20	18.640	0.000
(g) Na (cube root)	0.11			3	23.00	0.000	0.923
	0.14	+		4	17.60	5.380	0.063
	0.15		−0.01	4	14.50	8.460	0.013
	0.17	+	−0.01	5	−8.70	14.300	0.001

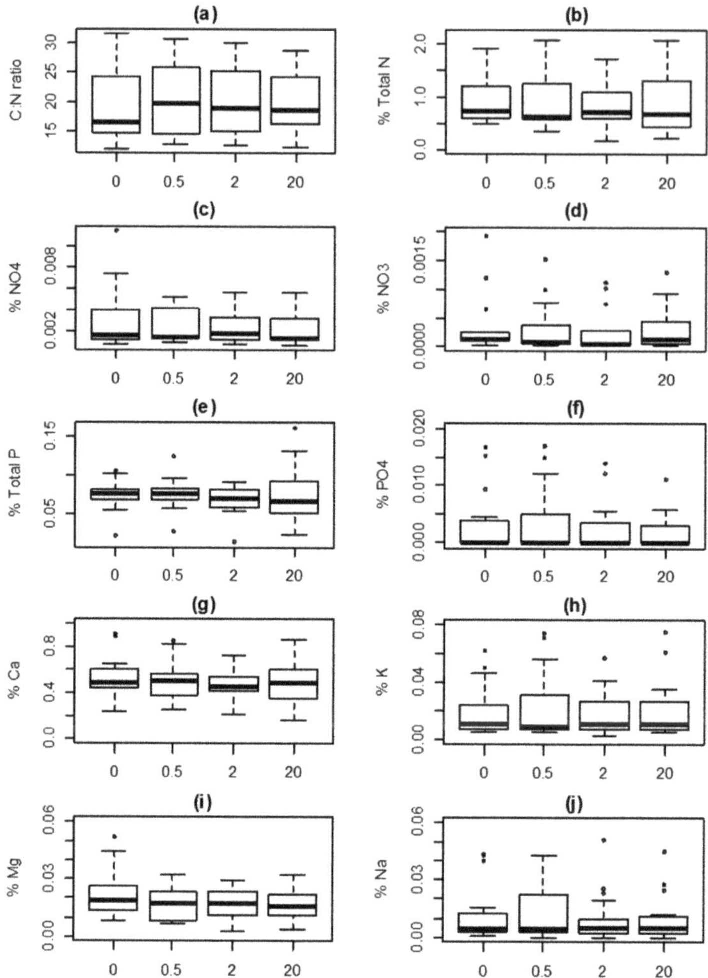

Figure 2. Response values in soil with increasing distance (0, 0.5, 2 and 20 m) from lynx-killed roe deer carcasses in southeastern Norway. (**a**) C/N ratio; (**b**) nitrogen, N; (**c**) ammonium, NH_4+; (**d**) nitrate, NO_3-; (**e**) phosphorus, P; (**f**) phosphate, PO_4-; (**g**) calcium, Ca; (**h**) potassium, K; (**i**) magnesium, Mg; (**j**) sodium, Na. All values are in percentage mass; the boxplots show the median (black line within the box) and quartiles (the boundaries of the box). Whiskers above and below the box indicate the 10 th and 90 th percentiles. Points above and below the whiskers indicate outliers outside the 10 th and 90 th percentiles.

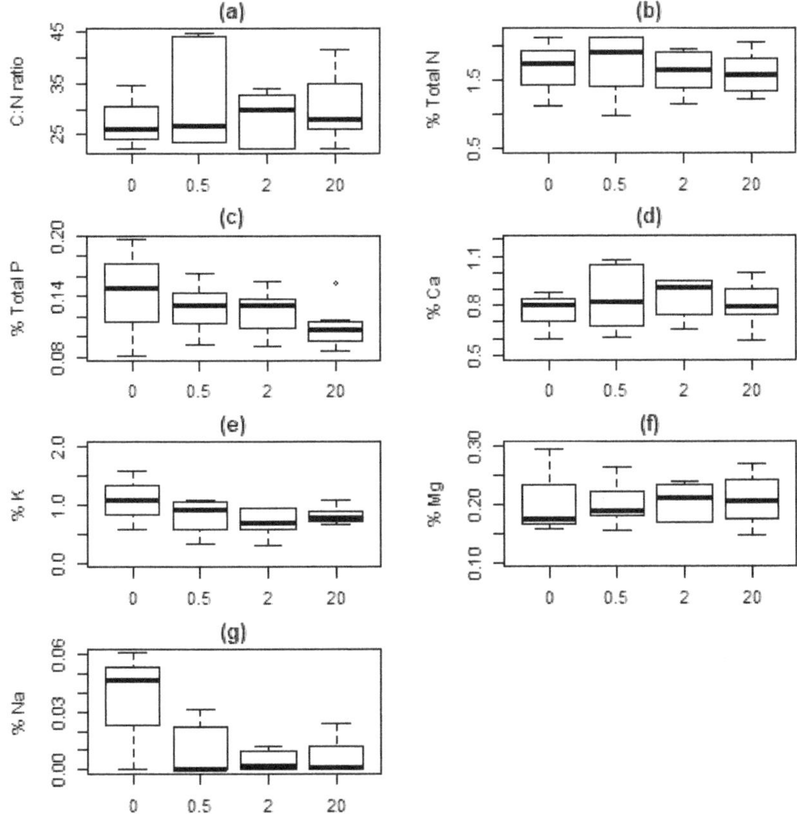

Figure 3. Response values in *Vaccinium myrtillus* with increasing distance (0, 0.5, 2 and 20 m) from lynx-killed roe deer carcasses in south-eastern Norway. (**a**) C:N ratio; (**b**) Nitrogen, N; (**c**) Phosphorus, P; (**d**) Calcium, Ca; (**e**) Potassium, K; (**f**) Magnesium, Mg; (**g**) Sodium, Na. All values are in percentage mass. For additional information, see Figure 2.

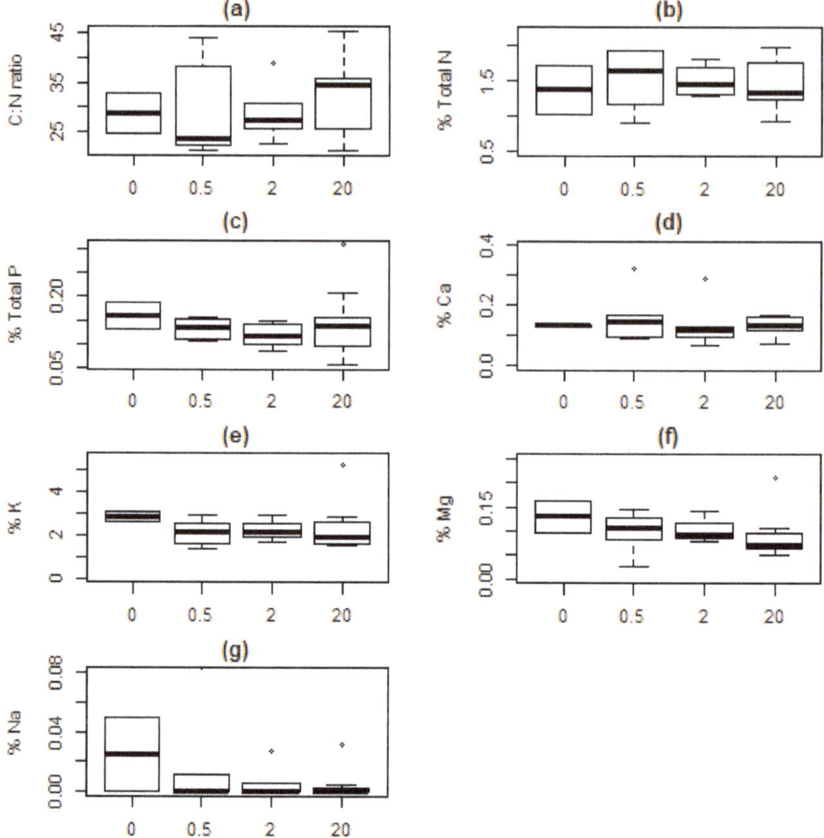

Figure 4. Response values in *Avenella flexuosa* with increasing distance (0, 0.5, 2 and 20 m) from lynx-killed roe deer carcasses in south-eastern Norway. (**a**) C:N ratio; (**b**) Nitrogen, N; (**c**) Phosphorus, P; (**d**) Calcium, Ca; (**e**) Potassium, K; (**f**) Magnesium, Mg; (**g**) Sodium, Na. All values are in percentage mass. For additional information, see Figure 2.

4. Discussion

In contrast to our predictions, we found no detectable below-ground impacts of lynx-killed roe deer carcasses. Our results differ from other published studies, such as on muskox (*Ovibos moschatus*) carcasses in the arctic [19], moose in Isle Royale [18], American bison (*Bison bison*) in prairies [21] and kangaroos (*Macropus giganteus*) in Eucalyptus woodland [42], that detected an effect of carcass decomposition on soil and vegetation nutrients concentration [18–21,25], begging the question, what differs in our study system? A number of potential explanations present themselves, but all must be seen in light of the fact that only the liquids resulting from the invertebrate maggot activity and microbial decomposition leach into the soil [25].

First, the effect of roe deer carcasses on soil and vegetation is expected to vary depending on how much of its prey is consumed by the lynx and how much the remnants are consumed or dispersed by scavengers [24,26]. Since roe deer are quite small ungulates, with a mean adult body mass of 20–30 kg [43], and lynx tend to consume most of the consumable meat and organs over the course of 1 to 5 days after a kill [44,45], the amount of nutrients available for leaking into the soil is rather limited. Moreover, vertebrate scavengers tend to arrive within a few hours or days and rapidly remove the rest of the meat/organs [30]. On our sample of kills, more than 90% of the meat was removed from almost all carcasses when we first examined the kill sites several days after the lynx had abandoned the kills.

This issue is enhanced by the fact that a good proportion of the kills (6 out of 18) happened in the winter season, when insects are inactive and decomposition rates are lower—thus giving even more time for scavengers to remove meat and disperse body parts. Combined, these issues point to a need to consider the methodology when comparing studies. Our use of real predator-killed and consumed carcasses contrasts to most other studies that simulate this effect with whole carcasses and which often exclude scavengers through fencing [42]. This difference needs to be considered when generalizing between carrion derived from different sources and between studies with different methodologies.

Second, another reason for not finding an effect could lie with the time interval between roe deer death and carcass sampling. In this study, similarly to a study conducted on much larger European bison (*Bison bonasus*) carcasses in a temperate forest [20], we collected samples two or more years after death. For European bison, we were able to detect an effect on pH and calcium that lasted up to seven years after the death of the animal. However, the turnover of nitrate in soil was very fast (up to one year) even though bison are significantly larger than roe deer (average body mass across sexes is over 500 kg [46]) and we might expect the impact on soil and vegetation to last longer. In a tallgrass prairie system, the nutrients released from American bison carcasses affected plant species composition for at least five years after bison death, thus increasing grassland heterogeneity [21]. Other studies on kangaroo carcasses (ca 30 kg) have also shown that nutrient effects can still be evident after five years [42].

Thirdly, the lack of elevation in nutrients in the plants that we sampled close to a carcass could be because sampled *V. myrtillus* belonged to the same clone and therefore to various degree exchange resources underground between ramets or store nutrients in storage organs [47–49], thus potentially diluting the spatial impact of the nutrient pulse. It is, however, unlikely that the controls (20 m) should belong to the same clone as the plants close to the carcass. A final consideration concerns our sample sizes. Although not very large, they are very much within the range of sample sizes used in similar studies that have found significant results. Therefore, although we may not be able to reject the existence of any effects of the carcasses, we feel comfortable to say that if they exist, they are much more subtle than other studies have shown for larger carcasses in different ecosystems.

Our findings underline the complexity, diversity and context dependence of ecological processes, as well as illuminating the dangers of over-generalizations from case studies to universal principles. Although there is a growing list of case studies documenting a diversity of large predator-mediated trophic cascades in terrestrial as well as aquatic ecosystems, there are also multiple reviews that underline that there are also many exceptions [6,50] and calls for caution [51]. Genuine scientific progress in this value-laden field will only be possible by building up a solid body of positive and negative results embracing the diversity of contexts (both ecological and anthropogenic) in which predation occurs in the Anthropocene [52,53].

Author Contributions: Conceptualization, I.J.M.T. and J.D.C.L.; methodology, C.M., C.S., I.J.M.T. and J.D.C.L.; formal analysis, C.M. and I.J.M.T.; writing—original draft preparation, C.M., C.S., I.J.M.T. and J.D.C.L.; writing—review and editing, C.M., C.S., I.J.M.T. and J.D.C.L.; funding acquisition, J.D.C.L. and I.J.M.T. All authors have read and agreed to the published version of the manuscript.

Funding: This study was conducted as part of the Scandlynx project which is mainly funded by the Norwegian Directorate for Nature Management, the Research Council of Norway (projects 134242, 165814, 183176, 212919, 251112), the county governor's office of Viken county, and the Norwegian Institute for Nature Research. IJMT was also funded by two individual grants from the Research Council of Norway's cultural agreement program with the Netherlands (2002-03 and 2003-04) and two grants from the Schure-Beijerinck-Popping fund. CM was funded by a postdoctoral grant at the Norwegian University for Science and Technology, Trondheim.

Acknowledgments: The authors are grateful to all the landowners who permitted our activity on their land, and to Martin Hauger and Øistein Høgseth for assistance in the field. The Ski municipality assisted by providing carcasses and the Norwegian University for Life Sciences provided drying ovens for vegetation samples.

Conflicts of Interest: The authors declare no conflict of interest.

References

1. Smith, D.W.; Peterson, R.O.; Houston, D.B. Yellowstone after wolves. *Bioscience* **2003**, *53*, 330–340. [CrossRef]
2. Berger, J.; Smith, D.W. Restoring functionality in Yellowstone with recovering carnivores: Gains and uncertainties. In *Large Carnivores and the Conservation of Biodiversity*; Ray, J.C., Redford, K.H., Steneck, R.S., Berger, J., Eds.; Island Press: Washington, DC, USA, 2005; pp. 100–109.
3. McShea, W.J. Forest ecosystems without carnivores: When ungulates rule the world. In *Large Carnivores and the Conservation of Biodiversity*; Ray, J.C., Redford, K.H., Steneck, R.S., Berger, J., Eds.; Island Press: Washington, DC, USA, 2005; pp. 138–153.
4. Ray, J.C. Large carnivorous animals as tools for conserving biodiversity: Assumptions and uncertainties. In *Large Carnivores and the Conservation of Biodiversity*; Ray, J.C., Redford, K.H., Steneck, R.S., Berger, J., Eds.; Island Press: Washington, DC, USA, 2005; pp. 34–56.
5. Soule, M.E.; Estes, J.A.; Miller, B.; Honnold, D.L. Strongly interacting species. conservation policy, management, and ethics. *Bioscience* **2005**, *55*, 168–176. [CrossRef]
6. Ray, J.C.; Redford, K.H.; Berger, J.; Steneck, R.S. Conclusion: Is large carnivore conservation equivalent to biodiversity conservation and how can we achieve both? In *Large Carnivores and the Conservation of Biodiversity*; Ray, J.C., Redford, K.H., Steneck, R.S., Berger, J., Eds.; Island Press: Washington, DC, USA, 2005; pp. 400–428.
7. Pace, M.L.; Cole, J.J.; Carpenter, S.R.; Kitchell, J.F. Trophic cascades revealed in diverse ecosystems. *Trends Ecol. Evol.* **1999**, *14*, 483–488. [CrossRef]
8. Schmitz, O.J.; Hamback, P.A.; Beckerman, A.P. Trophic cascades in terrestrial systems: A review of the effects of carnivore removals on plants. *Am. Nat.* **2000**, *155*, 141–153. [CrossRef]
9. Hobbs, N.T. Large herbivores as sources of disturbance in ecosystems. In *Large Herbivore Ecology, Ecosystem Dynamics and Conservation*; Pastor, J., Danell, K., Duncan, P., Bergström, R., Eds.; Cambridge University Press: Cambridge, MA, USA, 2006; pp. 261–288.
10. Ripple, W.J.; Beschta, R.L. Linking wolves and plants: Aldo Leopold on trophic cascades. *Bioscience* **2005**, *55*, 613–621. [CrossRef]
11. Berger, J.; Stacey, P.B.; Bellis, L.; Johnson, M.P. A mammalian predator-prey imbalance: Grizzly bear and wolf extinction affect avian neotropical migrants. *Ecol. Appl.* **2001**, *11*, 947–960. [CrossRef]
12. DeVault, T.L.; Rhodes, O.E.; Shivik, J.A. Scavenging by vertebrates: Behavioral, ecological, and evolutionary perspectives on an important energy transfer pathway in terrestrial ecosystems. *Oikos* **2003**, *102*, 225–234. [CrossRef]
13. Hunter, J.S.; Durant, S.M.; Caro, T.M. Patterns of scavenger arrival at cheetah kills in Serengeti National Park Tanzania. *Afr. J. Ecol.* **2007**, *45*, 275–281. [CrossRef]
14. Selva, N.; Fortuna, M.A. The nested structure of a scavenger community. *Proc. R. Soc. B Biol. Sci.* **2007**, *274*, 1101–1108. [CrossRef]
15. Selva, N.; Jedrzejewska, B.; Jedrzejewski, W.; Wajrak, A. Scavenging on European bison carcasses in Bialowieza Primeval Forest (eastern Poland). *Ecoscience* **2003**, *10*, 303–311. [CrossRef]
16. Selva, N.; Jedrzejewska, B.; Jedrzejewski, W.; Wajrak, A. Factors affecting carcass use by a guild of scavengers in European temperate woodland. *Can. J. Zool.* **2005**, *83*, 1590–1601. [CrossRef]
17. Wilmers, C.C.; Crabtree, R.L.; Smith, D.W.; Murphy, K.M.; Getz, W.M. Trophic facilitation by introduced top predators: Grey wolf subsidies to scavengers in Yellowstone National Park. *J. Anim. Ecol.* **2003**, *72*, 909–916. [CrossRef]
18. Bump, J.K.; Peterson, R.O.; Vucetich, J.A. Wolves modulate soil nutrient heterogeneity and foliar nitrogen by configuring the distribution of ungulate carcasses. *Ecology* **2009**, *90*, 3159–3167. [CrossRef] [PubMed]
19. Danell, K.; Berteaux, D.; Brathen, K.A. Effect of muskox carcasses on nitrogen concentration in tundra vegetation. *Arctic* **2002**, *55*, 389–392. [CrossRef]
20. Melis, C.; Selva, N.; Teurlings, I.; Skarpe, C.; Linnell, J.D.C.; Andersen, R. Soil and vegetation nutrient response to bison carcasses in Bialeowieza Primeval Forest, Poland. *Ecol. Res.* **2007**, *22*, 807–813. [CrossRef]
21. Towne, E.G. Prairie vegetation and soil nutrient responses to ungulate carcasses. *Oecologia* **2000**, *122*, 232–239. [CrossRef]
22. Schmitz, O.J.; Hawlena, D.; Trussell, G.C. Predator control of ecosystem nutrient dynamics. *Ecol. Lett.* **2010**, *13*, 1199–1209. [CrossRef] [PubMed]

23. Naiman, R.J.; Bilby, R.E.; Schindler, D.E.; Helfield, J.M. Pacific salmon, nutrients, and the dynamics of freshwater and riparian ecosystems. *Ecosystems* **2002**, *5*, 399–417. [CrossRef]
24. Putman, R.J. *Carrion and Dung: The Decomposition of Animal Waste*; Edward Arnold: London, UK, 1983.
25. Carter, D.O.; Yellowlees, D.; Tibbett, M. Cadaver decomposition in terrestrial ecosystems. *Naturwissenschaften* **2007**, *94*, 12–24. [CrossRef]
26. Barton, P.S.; Cunningham, S.A.; Lindenmayer, D.B.; Manning, A.D. The role of carrion in maintaining biodiversity and ecological processes in terrestrial ecosystems. *Oecologia* **2013**, *171*, 761–772. [CrossRef]
27. Wardle, D.A.; Bardgett, R.D.; Klironomos, J.N.; Setala, H.; van der Putten, W.H.; Wall, D.H. Ecological linkages between aboveground and belowground biota. *Science* **2004**, *304*, 1629–1633. [CrossRef] [PubMed]
28. Bump, J.K.; Webster, C.R.; Vucetich, J.A.; Peterson, R.O.; Shields, J.M.; Powers, M.D. Ungulate Carcasses Perforate Ecological Filters and Create Biogeochemical Hotspots in Forest Herbaceous Layers Allowing Trees a Competitive Advantage. *Ecosystems* **2009**, *12*, 996–1007. [CrossRef]
29. DeVault, T.L.; Brisbin, I.L.; Rhodes, O.E. Factors influencing the acquisition of rodent carrion by vertebrate scavengers and decomposers. *Can. J. Zool.* **2004**, *82*, 502–509. [CrossRef]
30. Ray, R.R.; Seibold, H.; Heurich, M. Invertebrates outcompete vertebrate facultative scavengers in simulated lynx kills in the Bavarian Forest National Park, Germany. *Anim. Biodiv. Conserv.* **2014**, *37*, 77–88.
31. Turner, K.L.; Abernethy, E.F.; Conner, L.M.; Olin, E.; Beasley, J.C. Abiotic and biotic factors modulate carrion fate and vertebrate scavenging communities. *Ecology* **2017**, *98*, 2413–2424. [CrossRef]
32. Nilsen, E.B.; Linnell, J.D.C.; Odden, J.; Andersen, R. Climate, season, and social status modulate the functional response of an efficient stalking predator: The Eurasian lynx. *J. Anim. Ecol.* **2009**, *78*, 741–751. [CrossRef]
33. Odden, J.; Linnell, J.D.C.; Andersen, R. Diet of Eurasian lynx, Lynx lynx, in the boreal forest of southeastern Norway: The relative importance of livestock and hares at low roe deer density. *Eur. J. Wildlife Res.* **2006**, *52*, 237–244. [CrossRef]
34. Novozamsky, I.; Houba, V.J.G.; Temminghoff, E.; van der Lee, J.J. Determination of 'total' N and 'total' P in a single soil digest. *Neth. J. Agric. Sci.* **1984**, *32*, 322–324.
35. Walinga, I.; Van Der Lee, J.J.; Houba, V.J.G.; Van Vark, W.; Novozamsky, I. Digestion in tubes with H2SO4-salicylic acid- H2O2 and selenium and determination of Ca, K, Mg, N, Na, P, Zn. In *Plant Analysis Manual*; Walinga, I., Van Der Lee, J.J., Houba, V.J.G., Van Vark, W., Novozamsky, I., Eds.; Springer: Dordrecht, The Netherlands, 1995; pp. 7–45.
36. Houba, V.J.G.; Temminghoff, E.J.M.; Gaikhorst, G.A.; van Vark, W. Soil analysis procedures using 0.01 M calcium chloride as extraction reagent. *Commun. Soil Sci. Plant Anal.* **2000**, *31*, 1299–1396. [CrossRef]
37. Pinheiro, J.C.; Bates, D.M. *Mixed-Effect Models in S and S-Plus*; Springer: New York, NY, USA, 2002.
38. Burnham, K.P.; Anderson, D.R. *Model Selection and Multi-Model Inference: A Practical Information-Theoretic Approach*; Springer: Berlin/Heidelberg, Germany, 2002.
39. Team, R.C. *R: A Language and Environment for Statistical Computing*; R Foundation for Statistical Computing: Vienna, Austria, 2013; Available online: https://www.r-project.org/ (accessed on 12 September 2020).
40. Bartoń, K. MuMIn: Multi-Model Inference; R Package Version 1.43.17; 2020. Available online: https://www.rdocumentation.org/packages/MuMIn/versions/1.43.17 (accessed on 12 September 2020).
41. Bates, D. lme4: Linear Mixed-Effects Models Using S4 Classes; 0.99875-6. 2007. Available online: https://cran.r-project.org/web/packages/lme4/index.html (accessed on 12 September 2020).
42. Barton, P.S.; McIntyre, S.; Evans, M.J.; Bump, J.K.; Cunningham, S.A.; Manning, A.D. Substantial long-term effects of carcass addition on soil and plants in a grassy eucalypt woodland. *Ecosphere* **2016**, *7*, e01537. [CrossRef]
43. Andersen, R.; Duncan, P.; Linnell, J.D.C. *The European Roe Deer: The Biology of Success*; Scandinavian University Press: Oslo, Norway, 1998; Volume 376.
44. Okarma, H.; Jedrzejewski, W.; Schmidt, K.; Kowalczyk, R.; Jedrzejewska, B. Predation of Eurasian lynx on roe deer and red deer in Bialowieza Primeval Forest, Poland. *Acta Theriol.* **1997**, *42*, 203–224. [CrossRef]
45. Breitenmoser, U.; Breitenmoser-Würsten, C. *Der Luchs: Ein Grossraubtier in der Kulturlandschaft*; Salm Verlag: Bern, Switzerland, 2008.
46. Krasinska, M.; Krasinski, Z.A. Body mass and measurements of the European bison during postnatal development. *Acta Theriol.* **2002**, *47*, 85–106. [CrossRef]
47. Klimešová, J.; de Bello, F. CLO-PLA: The database of clonal and bud bank traits of Central European flora. *J. Veg. Sci.* **2009**, *20*, 511–516. [CrossRef]

48. Klimešová, J.; Klimeš, L. Clo-Pla3—Database of Clonal Growth of Plants from Central Europe. 2006. Available online: https://clopla.butbn.cas.cz (accessed on 12 September 2020).

49. Tolvanen, A. Differences in Recovery between a Deciduous and an Evergreen Ericaceous Clonal Dwarf Shrub after Simulated Aboveground Herbivory and Belowground Damage. *Can. J. Bot.* **1994**, *72*, 853–859. [CrossRef]

50. Shurin, J.B.; Markel, R.W.; Matthews, B. Comparing trophic cascades across ecosystems. In *Trophic Cascades: Predators, Prey and the Changing Dynamics of Nature*; Terborgh, J., Estes, J.A., Eds.; Island Press: Washington, DC, USA, 2010; pp. 319–335.

51. Hayward, M.W.; Edwards, S.; Fancourt, B.A.; Linnell, J.D.C.; Nilsen, E.B. Top-down control of ecosystems and the case for rewilding: Does it all add up? In *Rewilding*; Pettorelli, N., Durant, S.M., du Toit, J.T., Eds.; Cambridge University Press: Cambridge, MA, USA, 2018; pp. 325–354.

52. Boitani, L.; Linnell, J.D.C. Bringing large mammals back: Large carnivores in Europe. In *Rewilding European Landscapes*; Pereira, H.M., Navarro, L.M., Eds.; Springer: Cham, Switzerland, 2015; pp. 67–84.

53. Linnell, J.D.C.; Promberger, C.; Boitani, L.; Swenson, J.E.; Breitenmoser, U.; Andersen, R. The linkage between conservation strategies for large carnivores and biodiversity: The view from the "half-full" forests of Europe. In *Carnivorous Animals and Biodiversity: Does Conserving One Save the Other?* Ray, J.C., Redford, K.H., Steneck, R.S., Berger, J., Eds.; Island Press: Washington, DC, USA, 2005; pp. 381–398.

Review

Predation and Scavenging in the City: A Review of Spatio-Temporal Trends in Research

Álvaro Luna [1], Pedro Romero-Vidal [2,3] and Eneko Arrondo [2,4,*]

[1] Asociación Brutal, Calle Cuna, 16, Primera Planta, 41004 Sevilla, Spain; alvalufer@gmail.com
[2] Department of Conservation Biology, Estación Biológica de Doñana—CSIC, 41004 Sevilla, Spain; pedroromerovidal123@gmail.com
[3] Department of Physical, Chemical and Natural Systems, Universidad Pablo de Olavide, 41004 Sevilla, Spain
[4] Department of Applied Biology, Universidad Miguel Hernández, 03202 Elche, Spain
* Correspondence: bioeaf@gmail.com

Abstract: Many researchers highlight the role of urban ecology in a rapidly urbanizing world. Despite the ecological and conservation implications relating to carnivores in cities, our general understanding of their potential role in urban food webs lacks synthesis. In this paper, we reviewed the scientific literature on urban carnivores with the aim of identifying major biases in this topic of research. In particular, we explored the number of articles dealing with predation and scavenging, and assessed the geographical distribution, biomes and habitats represented in the scientific literature, together with the richness of species reported and their traits. Our results confirmed that scavenging is largely overlooked compared to predation in urban carnivore research. Moreover, research was biased towards cities located in temperate biomes, while tropical regions were less well-represented, a pattern that was more evident in the case of articles on scavenging. The species reported in both predation and scavenging articles were mainly wild and domestic mammals with high meat-based diets and nocturnal habits, and the majority of the studies were conducted in the interior zone of cities compared to peri-urban areas. Understanding the trophic role of carnivores in urban environments and its ecological consequences will require full recognition of both their predation and scavenging facets, which is especially desirable given the urban sprawl that has been predicted in the coming decades.

Keywords: anthropogenic food; diet; urban habitats; ecological functions; carnivorous

Citation: Luna, Á.; Romero-Vidal, P.; Arrondo, E. Predation and Scavenging in the City: A Review of Spatio-Temporal Trends in Research. *Diversity* **2021**, *13*, 46. https://doi.org/10.3390/d13020046

Academic Editor: Luc Legal
Received: 22 October 2020
Accepted: 28 December 2020
Published: 25 January 2021

Publisher's Note: MDPI stays neutral with regard to jurisdictional claims in published maps and institutional affiliations.

1. Introduction

Today, in an increasingly urbanized world, scientists, conservationists, and politicians agree that understanding the patterns that explain the biodiversity of cities and conserving this biodiversity along with its ecological functions is a priority within urban planning and nature conservation [1–4]. Typically, the growth of cities has been linked to biodiversity loss [5–7]. However, recent studies have demonstrated that cities are exploited by more species than previously thought (sometimes reaching higher population densities than in their original habitats [2,8]), and are even capable of hosting endangered species [9–12]. Factors related to cities that influence species composition and their exploitation of this novel ecosystem include urban structure and development [6,13,14], different types of pollution found in cities (e.g., noise [15], artificial light [16], chemical contamination [17]), and the existence of refuge and food sources [18]. For vertebrates such as birds, urbanization acts as a filter, but a combination of traits, including phenotypic and behavioral flexibility, dispersal strategies, and niche flexibility, allow a certain number of species to exploit cities [19–22]. The same can be attributed to the other most common vertebrates in cities, mammals, as certain reproduction-related traits and their behavioral flexibility favor the successful use of cities for a limited number of taxa [23,24]. Other factors influencing urban species composition include diet-related aspects, as species that thrive in cities usually can feed on a wide variety of foods, which sometimes are directly or indirectly related to humans [19,25]. Meat is an abundant and heterogeneously distributed food resource present in urban and peri-urban areas [22,26–28]. It appears in the form of both domestic and wild vertebrates, and also as carrion (i.e., roadkill and anthropogenic refuse [29,30]). This mostly occurs in cities with inefficient waste treatment, but also in landfills and ports located in typical peri-urban areas [26–28]. The diet of several urban carnivores has been assessed, and as such we have gained an appreciation both for the items they consume and the prevalence of the different sources of food. Thus, different prey form part of the urban diet of raptors, mammals, and other carnivores [31,32]. Moreover, a growing amount of research reports the progressive inclusion of human refuse in the diet of urban animals in the form of organic and inorganic garbage [33–35]. In general, urban carnivores range from obligate carnivores, such as the barn owls (*Tyto alba*) [31], to generalist omnivores, such as the red fox (*Vulpes vulpes*) [36] and corvids [37], which eat meat, vegetables, fruits, and berries, with different preponderance of each food source depending on the species. Therefore, urban carnivorous species can consume meat (i.e., vertebrate biomass) in two different ways: by predation (actively hunting their prey) [32,38,39] or by scavenging (eating the remains of already dead animals or garbage) [27,40]. Between these two extremes, several species act as facultative scavengers, consuming both live prey and carrion or human refuse, such as the coyotes (*Canis latrans*) [41] and gulls [42].

The feeding behavior observed by carnivores in cities contributes to the structure of natural urban communities, for example, by limiting the population size of their prey [43]. In some cases, they can provide benefits to humans, especially when they prey upon certain synanthropic species (e.g., feral pigeons [44] or rodents [31]). An example is found in Indian cities, where the consumption of stray dogs by leopards is expected to reduce disease transmission [45]. However, the constant availability of refuse can also alter the diet preferences of carnivores in anthropized areas, affecting the trophic dynamics and population densities of urban carnivores and their potential prey [29,30,46–48]. Regarding scavengers, they play additional and concrete roles in cities, by contributing to accelerating the process of carrion decomposition [49,50]. Moreover, the consumption of subsided food and the high abundance of prey in cities can have a profound effect at the individual level, influencing the growth rate, body condition, and survival of carnivores. In addition, there are also obvious negative consequences, for example, when feeding on resources with low nutritional value or indigestible waste [30,51]. Furthermore, carnivores in cities are also confronted with new threats, common to the urban environment, such as an increased probability of collision with buildings or vehicles [27,52,53]. Attending to this, cities may become ecological traps for certain species [54–56]. However, although mostly unstudied, urban exploitation by carnivores may play a potential role in their conservation, as observed for some scavengers feeding mostly on human subsided food [57,58].

In recent years, different reviews have analyzed the knowledge generated within urban ecology, and have proposed how this discipline can advance along multiple research directions [59,60]. Focusing on urban biodiversity studies, McPhearson et al. [61], established the acquisition of a better understanding of how the ecological community structure, including invasive species, affects ecosystem dynamics, structure, and function in cities as a future priority. Moreover, Magle et al. [60] revealed how urban ecology research is biased towards birds and mammals, is generally focused on behavior and conservation issues, and is carried out to a lesser extent in less developed countries. Apparently, studies focused on the diet of carnivores, especially scavengers, seem to have been of little priority, despite the importance they may have for understanding urban wildlife communities, the ecosystem services they provide, and the potential conservation implications.

In this study, we aim to review the scientific literature on urban carnivores, distinguishing between the predatory and scavenging function of this ecological group, to offer an integrative first approach about the spatio-temporal trends observed in the scientific literature focused on this topic. Concerning the information contained in the articles, we also explored in which locations within the urban matrix the studies were conducted (the interior of the city vs. the peri-urban areas), and, in the case of scavengers, we analyzed if the studies were carried out based on natural conditions or developing experiments with carrion provided by the researchers. Moreover, we explored the diversity of carnivores reported in the reviewed articles, together with the primary physical and behavioral traits that may contribute to explaining their presence in the reviewed studies. Considering that scavenger impact on ecology has been little studied until recent decades [62], we hypothesize that most of the published research concerning carnivores' diet and presence in cities has been devoted more to predators than to scavengers. In addition, following the suggestions from Magle et al. [60] about the current bias in urban ecology research on developed countries, we try to confirm that this trend also occurs in the case of carnivorous studies in cities. Additionally, we consider that the carnivores and scavenger richness obtained in our review would be higher, with many species recorded in studies from Africa, Latin America and Asia. To test this hypothesis, the greater species richness of tropical regions in which most of the developing countries are located [63,64] must be considered, as this logically leads to greater chances of more species potentially exploiting cities. Moreover, the poorer waste management of developing countries offers greater possibilities for feeding for scavengers in the form of all types of refuse, contributing to the presence of different scavengers. Lastly, attending to the characteristics of the vertebrates that most commonly exploit cities, we predict that the most represented predators and scavengers will be synanthropic species, mainly mammals and birds, with nocturnal habits that would permit these species to avoid potential conflict with humans.

2. Materials and Methods

2.1. Systematic Literature Review

To review the scientific literature published about urban carnivores we followed the guidelines proposed by Haddaway et al. [65] and Lozano et al. [66]. First, we studied peer-reviewed articles published in journals available on Web of Science and Scopus databases. The search was applied to the title, without selecting any date range. We considered the inclusion of the grey literature, taking into account the contribution of these additional sources for systematic reviews [67,68]. However, in our searches, we did not find these technical reports, conferences and similar publications, and so we only included scientific articles. To avoid the heterogeneity inherent to different observers, only one of the authors searched the databases, by using a search string that combined different terms related to predation and scavenging in urban environments. We conducted two different searches including the terms (urban* OR city*) AND (scaveng* OR predat*). Second, we completed our database, both in the case of urban predation and urban scavenging, by applying a "snowball" procedure, including additional articles found in the previously selected reference lists [66]. Third, we completed the dataset on predation and scavenging by including more articles found in a complementary non-systematic search, based mainly on Google Scholar.

We explored the content of the compiled articles with a two-step process. First, we screened titles and abstracts to ensure we only included articles focused on the urban predation and scavenging diet of urban wildlife. Accordingly, studies located in human dominated but non-urban sites, for example, those conducted in transformed landscapes such as agricultural areas, were excluded. Moreover, we also left out those nature reserves located in other anthropized locations with less urban development. Equally, articles focused on human scavenging in garbage dumps (picking through garbage for food scraps), and those referring to forensic science were discarded. Secondly, we read the main text of the articles to see how and to what extent urban predation/scavenging was addressed. We considered a species to be an urban predator when it was recorded in urban environments actively preying on vertebrates (including eggs) while we considered a species to be an urban scavenger when it was recorded consuming carrion (including human waste). Regarding the topic in the studies we reviewed, in our final list, we discarded those studies based on scavenger species but exploring other aspects of their biology and ecology different from their diet, as well as non-empirical studies (i.e., reviews, conceptual papers and anecdotal short notes without detailed data to use in further analyses).

2.2. Data Extraction

From the retained articles, we extracted the following information: (i) year of publication; (ii) study site, including the biome (based on the classification used by Olson et al. [69] to define major habitat types), continent, country and city studied; (iii) population size of the city during the year of the study; (iv) sector of the city considered, classifying the articles as urban, (those conducted in the interior of the cities), and peri-urban (i.e., peripheral landfills and urban nature reserves surrounding cities, and residential neighborhoods); (v) studied taxa, at class, order and species level; (vi) richness of predator/scavenging species (this was calculated only in those studies that did not focus on the activity of a single species). Then, for each reported species, we identified (vii) the carnivore level, calculated as the proportion of diet based on vertebrates (including eggs). To calculate this, in the case of mammals and birds, we summed all dietary variables which included vertebrates, from Wilman et al. [70], (viii) average body mass of the species in grams and (ix) daily activity (mainly diurnal or nocturnal). For reptiles, we used complementary dietary bibliography [71–75]. Solely for scavengers, we included whether the study was conducted thinking specifically about scavenging or, in contrast, if this was merely an aspect within studies focused on diet in a broader sense. Additionally, when the studies were focused on scavenging, we added information regarding the origin of the carrion, to clarify whether the species was purposely fed with certain carrion for research purposes (fish heads, bird carcasses from domestic or wild animals, garbage) instead of consuming the refuse scattered in the city.

2.3. Statistical Analyses

We conducted General Linear Models (GLMs; Poisson error distribution and log link function) to explore the characteristics of the species that were most frequently studied in urban predation and urban scavenging studies. In order to do this, we only considered those studies that had analyzed the activity of one or more species within the city or surrounding area. We created two sets of models separating information about urban predation and urban scavenging studies. In both sets, the explanatory variables were the number of times each species appeared in the articles found during the bibliographic review. As explanatory variables we included: (a) taxonomic class; (b) taxonomic order; (c) carnivory level; (d) body mass; (e) daily rhythm. In both sets of models, we explored all combinations of the variables, considering only additive effects. To discuss the results, we followed an Information Criterion Approach. We selected the most parsimonious models (i.e., lowest AICc). In addition, we selected all models with a difference of ΔAICc < 2 compared to the most parsimonious model [76]. As the information criterion does not inform about the absolute predictive capacity of the models, we later discarded those selected models with little-relevant explanatory variables (i.e., the values at the 85% confidence interval which overlapped with 0 [76]. We used an 85% confidence interval to avoid the possible loss of valid models by using more restrictive and conservative approaches such as the 95% confidence interval [77]. We explored the overdispersion of selected models using the AER package [78]. Finally, to analyze how well our models fit, we calculated the explained deviance (D^2). All statistical analyses were performed in the R v.3.6.1 statistical platform [79].

3. Results

The results using predat* as the search term returned a total of 32 items, while using scaveng* offered only five results. Including only the articles found in the enclosed review, we observed that most predation articles were published after the 2000s, and all scavenging articles had been published in the last 10 years, with scavenging remaining in the background in the studies focused on carnivores in cities, although proving to be a nascent interest as a study topic (Figure 1). By adding the articles obtained with complementary searches (snowball and other sources), we increased the number of studies, obtaining a final dataset of 95 articles for urban predation and 47 articles for urban scavenging (see Tables S1–S4 for details). Regarding the articles focused on urban scavenging, at least 13 of those included in the final list clearly demonstrated that these studies were carried out under experimental conditions, providing the scavengers with carrion (mainly farm animals) instead of evaluating the consumption of refuse scattered in cities.

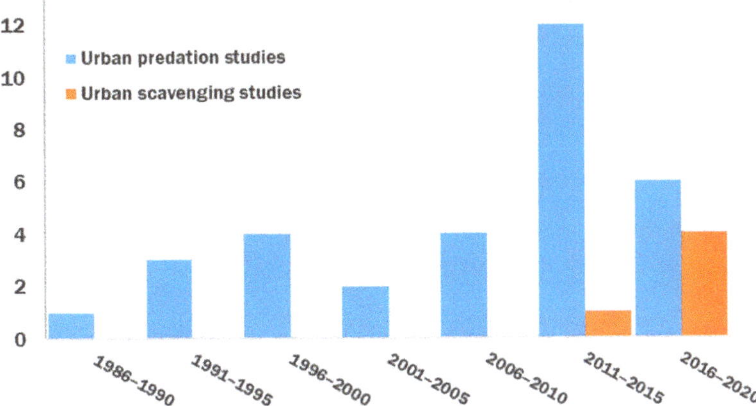

Figure 1. Temporal comparison between urban predation and urban scavenging published studies, based on the total of articles retained strictly from the systematic review, conducted with closed search.

We found that most of the studies analyzed were developed in temperate biomes (68.4% of the predation studies and 50% of the scavenging studies), followed by urban predation studies in Mediterranean forests (11.58%) and by tropical biomes (22.92%) in urban scavenging studies. Regarding the geographical locations, in the case of urban predation, 35 studies were performed in Europe, 32 in America, 17 in Oceania, 6 in Asia and 5 in Africa, with the United States of America ($n = 20$) and Australia ($n = 13$) accumulating the highest number of studies (Figure 2A, S1). On the other hand, urban scavenging studies were carried out in America in 12 cases, 10 in Europe, 9 in Oceania and Asia, and 8 in Africa. As for urban scavenging studies, Australia ($n = 9$) and the United Kingdom ($n = 5$) were the countries with the most published papers (Figure 2B, S2). The population size of the cities studied ranged from 3.54 to 9.1 million for predation studies and 3500 to 9 million for scavenging studies, which produced no significant difference ($t = 1.71$, df = 44.46, p-value = 0.095). Regarding the location where the studies were focused, we detected that the majority were conducted in the interior of cities or included both urban and peri-urban areas (Figure 3).

Of the 95 urban predation studies, 93 contained information on at least one species (the remaining two were focused on eggs predation without identifying the predators). In total, we recorded 100 species in the reviewed urban predation studies (62 birds, 28 mammals and 10 reptiles; see Table 1 for a complete list). The most represented orders were Passeriformes ($n = 23$), Accipitriformes ($n = 16$) and Carnivores ($n = 16$). Seven species appeared as predators in five or more articles: *Felis catus* ($n = 21$), *Falco tinnunculus* ($n = 8$), *Tyto alba* ($n = 8$), *Vulpes vulpes* ($n = 8$), *Canis latrans* ($n = 6$), *Canis lupus familiaris* ($n = 5$), and *Falco peregrinus* ($n = 5$) (Figure 2A). Regarding urban scavenging, we registered 49 species (25 birds, 21 mammals, and 3 reptiles) (Table 2), with Carnivores being the most represented group ($n = 11$), followed by Passerines ($n = 10$) and Accipitriformes ($n = 9$). The number of species is lower in comparison with those studies focused on predation, a logical result considering the smaller number of studies found on scavenging. All studies on scavenging (47) contained information on one or more species. Urban scavengers appearing in five or more different articles, were *Vulpes vulpes* (n = 11), and two domestic mammals, *Felis catus* (n = 9) and *Canis lupus familiaris* (n = 7) (Figure 2B). From the total of articles reviewed on predation, 21 contained sufficient information to calculate species richness (mean and standard deviation: 5.36 ± 5.26; range: 2–20 species per study), while 9 of the scavenging studies contained such information (mean and standard deviation: 7.00 ± 2.54; range: 3–10 species per study).

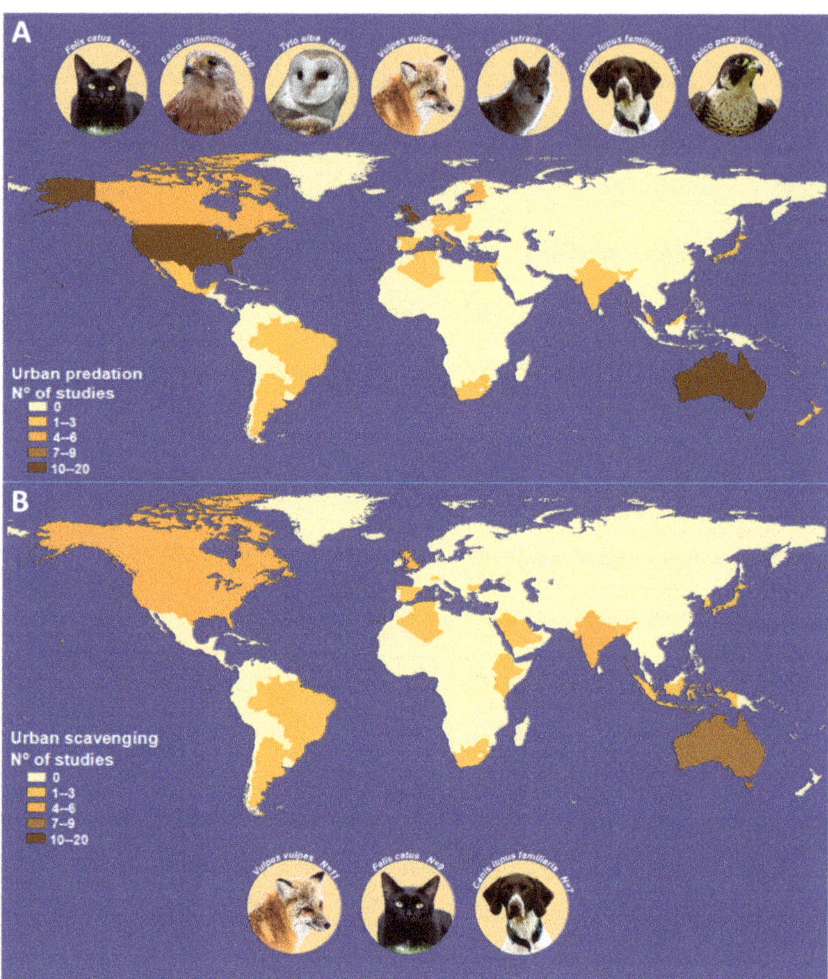

Figure 2. Number of urban predation (**A**) and scavenging (**B**) studies per country. Darker colors show high number of articles found for a given country, ranging from 0 to 20. The circles represent the species most reported in the articles (appearing on 5 or more occasions), ordered from most-reported to least-reported, from the top left corner. Pictures by: Manfredrichter, Ruthmcd, Miles, Wilda3, Skeeze and Ebor.

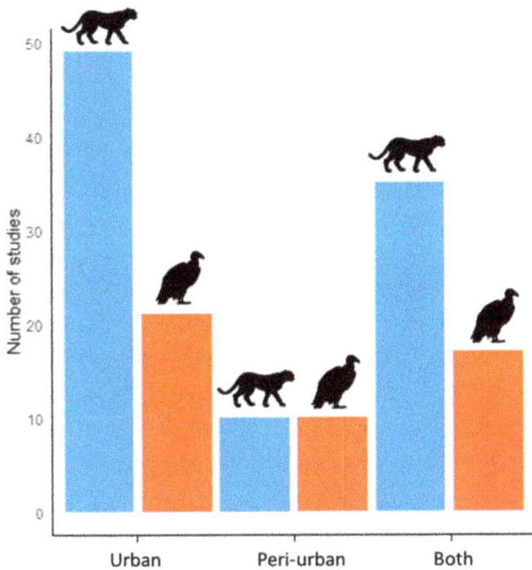

Figure 3. Number of studies conducted in urban and/or peri-urban areas on predation (blue bars) and scavenging (orange bars).

Table 1. Complete list of predators recorded in all articles reviewed. N is the number of studies that include each species.

Species	N	Species	N	Species	N
Felis catus	21	*Buteo lineatus*	1	*Sciurus carolinensis*	1
Falco tinnunculus	8	*Buteo platypterus*	1	*Sciurus niger*	1
Tyto alba	8	*Dacelo novaeguineae*	1	*Sciurus vulgaris*	1
Vulpes vulpes	8	*Dumetella carolinensis*	1	*Tamias striatus*	1
Canis familiaris	5	*Falco columbarius*	1	*Tamiasciurus hudsonicus*	1
Canis latrans	6	*Falco sparverius*	1	*Vulpes macrotis mutica*	1
Falco peregrinus	5	*Gralina cyanoleuca*	1	*Chaetophractus villosus*	1
Accipiter cooperii	4	*Gymnorhina tibicen*	1	*Geranoetus polyosoma*	1
Strix aluco	4	*Ictinia plumbea*	1	*Athene brama*	1
Martes foina	4	*Larus delawarensis*	1	*Athene cunicularia*	1
Rattus rattus	4	*Larus marinus*	1	*Bubo ascalaphus*	1
Corvus brachyrhynchos	3	*Larus michahellis*	1	*Bubo virginianus*	1
Cyanocitta cristata	3	*Leucophaeus atricilla*	1	*Strix varia*	1
Asio otus	3	*Manorina inelanocephala*	1	*Conepatus chinga*	1
Athene noctua	3	*Passer domesticus*	1	*Pseudalopex gymnocercus*	1
Procyon lotor	3	*Pica pica*	1	*Antechinus stuartii*	1
Canis lupus dingo	2	*Quiscalus quiscula*	1	*Genetta tigrina*	1
Accipiter gentilis	2	*Stephanoaetus coronatus*	1	*Glaucomys volans*	1
Falco femoralis	2	*Sturmis vulgaris*	1	*Mephitis mephitis*	1
Larus argentatus	2	*Troglodytes aedon*	1	*Mustela itatsi*	1
Molothrus ater	2	*Zosterops lateralis*	1	*Puma concolor*	1
Rupornis magnirostris	2	*Corvus cornix*	1	*Dasypus hybridus*	1
Strepera graculina	2	*Chroicocephalus ridibundus*	1	*Leopardus colocolo*	1
Ninox strenua	2	*Corvus corax*	1	*Galictis cuja*	1
Didelphis virginiana	2	*Perisoreus infaustus*	1	*Pantherophis emoryi*	1
Pseudocheirus peregrinus	2	*Garrulus glandarius*	1	*Philodryas olfersii*	1
Elaphe obsoleta lindheimerii	2	*Dendrocopos major*	1	*Phrynops geoffroanus*	1
Accipiter nisus	1	*Corvus corone*	1	*Psammophilus dorsalis*	1
Acridotheres tristis	1	*Corvus frugilegus*	1	*Pseudonaja affinis*	1
Aphelocoma californica	1	*Corvus monedula*	1	*Thamnophis sirtalis*	1
Aquila verreauxii	1	*Caracara plancus*	1	*Paraphimophis rusticus*	1
Buteo jamaicensis	1	*Circus buffoni*	1	*Erythrolamprus poecilogyrus*	1
Geranoaetus melanoleucus	1	*Circus cinereus*	1	*Xenodon dorbignyi*	1
Parabuteo unicinctus	1				

Table 2. Complete list of scavengers recorded in all articles reviewed. N is the number of studies that include each species.

Species	N	Species	N	Species	N
Vulpes vulpes	11	*Cyanocitta cristata*	1	*Haliastur sphenurus*	2
Felis catus	9	*Didelphis virginiana*	1	*Procyon lotor*	2
Canis lupus familiaris	7	*Erinaceus europaeus*	1	*Varanus varius*	2
Corvus orru	4	*Geococcyx californianus*	1	*Ciconia ciconia*	1
Milvus milvus	2	*Haliaeetus leucocephalus*	1	*Corvus coronoides*	1
Canis latrans	3	*Lanius ludovicianus*	1	*Corvus macrorhynchos*	1
Corvus corone	3	*Laridae* spp.	1	*Corvus splendens*	1
Crocuta crocuta	3	*Larus dominicanus*	1	*Varanus bengalensis*	1
Cynictis penicillata	3	*Larus michahellis*	1	*Varanus salvator*	1
Pica pica	3	*Larus novaehollandiae*	1	*Pica hudsonia*	1
Rattus spp.	3	*Larus pacificus*	1	*Scincidae* spp.	1
Chroicocephalus novaehollandiae	2	*Lutreolina crassicaudata*	1	*Scirus vulgaris*	1
Coragyps atratus	2	*Martes foina*	1	*Sciurus carolinensis*	1
Corvus brachyrhynchos	2	*Mephitis mephitis*	1	*Sciurus niger*	1
Galerella sanguinea	2	*Mustela putorius*	1	*Sus scrofa*	1
Haliaeetus leucogaster	2	*Necrosyrtes monachus*	1	*Sylvilagus floridanus*	1
Haliastur indus	2	*Neophron percnopterus*	1	*Tamias striatus*	1
Milvus migrans	2	*Terrapene carolina*	1		

In urban predation and scavenging studies, the most recorded species were those with meat-based diets and nocturnal activity (Table 3). In both sets, there were other models that included the explanatory variables carnivorous level and class (Table 3), at only a few tenths above the threshold of 2 ΔAICc. The selected models explained an important fraction of the deviance (0.27 in the case of the urban predation model and 0.29 in the case of the urban scavenging model) (Table 4). Both models showed very low values of over-dispersion (0.63 in the case of the urban predation model and 0.18 in the case of the urban scavenging model).

Table 3. AIC-based model selection to assess the characteristics of the scavenger species most recorded. Only models with informative variables are included. Model selected is represented in bold. Number of estimated parameters (K), AICc values, AICc differences (ΔAICc) compared to the highest ranked model (i.e., the one with the lowest AICc), Akaike weights (AICcWt), Cumulative weight (CumWt) and the variability of the models explained by the predictors (D^2) are represented.

Response Variable	Models	K	AICc	Delta_AICc	AICcWt	Cum.Wt	D^2
	Carnivore level + Daily rhythm	**3**	**347.804**	**0.000**	**0.798**	**0.798**	0.27
	Carnivore level + Class	4	350.596	2.792	0.198	0.996	
	Daily rhythm	2	358.698	10.893	0.003	0.999	
Urban predation	Carnivore level + Order + Daily rhythm	17	362.022	14.217	0.001	1.000	
	Carnivore level + Order	16	365.539	17.735	<0.001	1.000	
	Carnivore level	2	368.716	20.912	<0.001	1.000	
	Order	15	376.378	28.574	<0.001	1.000	
	Class	3	377.075	29.271	<0.001	1.000	
	NULL	1	387.227	39.423	<0.001	1.000	
	Carnivore level + Daily rhythm	**3**	**156.082**	**0.000**	**1.000**	**0.802**	0.29
	Carnivore level + Class	4	159.505	3.423	0.181	0.145	
	Daily rhythm	2	161.964	5.882	0.053	0.042	
Urban scavenging	Carnivore level + Order	2	164.850	8.768	0.012	0.010	
	Carnivore level	1	171.257	15.176	0.001	<0.001	
	Order	3	172.044	15.962	<0.001	<0.001	
	Class	18	203.664	47.582	<0.001	<0.001	
	NULL	19	209.212	53.131	<0.001	<0.001	

Table 4. Results of the General Linear Models (Poisson error distribution and identity linkage) selected to explain the number of studies in which each species appears. The estimate of the parameters (including the sign), the standard error of the parameters, the degrees of freedom of the models (DF) and variability explained by the models are represented (D^2).

Models	Explanatory Variables	Estimate	Std. Error	DF	D^2
Urban predation	(Intercept)	−0.079	0.161	98	0.27
	Carnivore level	0.007	0.002		
	Nocturnal	0.719	0.148		
Urban scavenging	(Intercept)	−0.018	0.275	42	0.29
	Carnivore level	0.010	0.003		
	Nocturnal	0.612	0.207		

4. Discussion

4.1. Temporal Trends of Predation/Scavenging Studies in Urban Ecology

In this review, we explored the scientific literature focused on urban predators and scavengers. A positive observation is that there is a growing interest in these subject as a study topic, as the articles have been increasing in number over the years. To strengthen the result that the topic is gaining interest, it can be noted that most of the articles reviewed had predation and scavenging as their main subject, and they did not relegate this aspect to a secondary interest within other subjects. Moreover, digging further into the general information assessed in the articles published, we can observe that most of the research studied not only peri-urban areas but also the interior of the cities, showing a clearer understanding of the activities of predators and scavengers in the cities themselves [80,81]. However, it is also true that in the case of scavenging studies, a non-negligible amount was carried out under experimental conditions [49,50,82,83], so it would be of interest to devote more effort to understanding the interaction of certain species with actual available refuse. When explaining the reason behind the temporal growth in the number of studies on predation and scavenging in cities, we consider that the general growth of urban ecology, as a relatively new discipline, may be also responsible for this positive trend [59,84]. Nevertheless, the gap detected between the number of articles referring to predators and scavengers reveals that less attention was devoted to urban scavenging, confirming that this topic has been widely overlooked and, to date, is lesser known and understood. To no surprise, our bibliographic search on urban scavenging only returned five articles, and despite having completed the dataset through the "snowball" and the complementary search, the total number of articles referring to urban scavenging was half of that found for urban predation. In any case, we cannot ignore that, in general, studies focused on scavenging ecology are fewer, with an increase only in recent years [62], a reality which may help to explain our result. Clearly, our search on both subjects (e.g., urban predators and scavengers) does not include the total studies on both subjects; however, the review we conducted shows that the number of studies is not related to the relevance that scavenging studies may acquire, attending to their ecological importance and the ecosystem services they might provide in cities [62,85].

4.2. The Bias Extant in Urban Predation and Scavenging Studies across Regions

We confirmed a geographical bias in the reviewed studies across different regions of the world [60], as the information is concentrated in a few specific countries while it remains scarce for other vast regions, especially the less developed ones (see Figure 2). We must address the simple fact that the lower scientific research activity detected in developing countries [86] may contribute to explaining the scarcity of articles focused on predation and scavenging in cities located in less developed countries. Moreover, we observed that a great variety of biomes are clearly underrepresented, even in regions where studies have been conducted, confirming the existence of a second spatial gap. Curiously, the regions which are less represented in the reviewed articles (mainly in Africa, Asia and Latin America) have urban areas that accumulate greater quantities of garbage and carrion in the streets (their poor management systems prevent them from handling the enormous amount of waste produced [87]), and thus are potentially suitable areas to study the interaction between wildlife and refuse. Additionally, the sites with worst waste collection are sites with potentially higher populations of synanthropic species (such as rats [88]) and semi-domestic animals roaming in the streets, and are thus locations where, hypothetically, more carnivores can thrive [45,89].

4.3. Predator and Scavenger Species in Urban Environments

Considering all the articles reviewed, we recorded a total of 100 predator species (found in 95 studies) and 49 scavengers (found in 47 studies). Some of these species are repeated, and appear in both sets of considered articles, given their opportunistic feeding habits that can lead them to both hunt and scavenge. Despite the greater number of species reported as predators, this may not necessarily be due to a greater presence of predators in cities, but rather to the fact that we only found around half the number of studies on scavenging compared to those on predation, thus reducing the possibility of detecting more species. This reinforces the need to increase research devoted to scavenging in cities. Life history, and ecological and behavioral traits predispose certain species to exploit urban environments [20,24,90,91], and the development of cities and associated impacts can limit or facilitate the successful exploitation of them by wildlife [92,93]. The reviewed studies do not offer sufficient data to analyze the influence of the elements associated with urbanization (i.e., noise, artificial light or human activities) on carnivorous occurrence. However, data extracted from the studies with ample information regarding predator and scavenger richness in cities suggest that the carnivore guild is reduced in urban environments, with an important range of variability. Further research is needed to gain knowledge of how certain characteristics inherent to the structure and environment of cities influence the occurrence of more carnivores. With regard to this, we suggest the inclusion of variables related to noise and acoustic pollution in studies related to urban carnivores, which are practically absent in the reviewed articles, despite their recognized impact on species exploiting cities [15–17]. In any case, we cannot forget that urban areas are normally characterized by lower biodiversity than wildlands [2,5–7], so it is expected that not all of the potential species inhabiting the surrounding areas of each city will be able to successfully exploit these human-dominated ecosystems.

Additionally, our results showed that the most encountered species are those with a highly meat-based diet, and, consequently, less consumption of vegetables or insects. Urban predators find a high availability of food associated with human environments (i.e., pets, pigeons and rats), which aids them in seeing that their trophic needs are easily met [36,94,95]. On the other hand, garbage constitutes an accessible resource which not only attracts obligate scavengers such as black vultures (*Coragyps atratus*) [30], but also seems to serve as a complement to the predators settled in the city. This is corroborated by the fact that most of the species detected in the scavenging studies also appeared on the predator list (Figure 2 and Tables 1 and 2). Regarding other traits pertaining to the carnivores recorded in the studies, most of the reported species showed nocturnal habits, coinciding with hours of less human activity. Thus, our results suggest that urban carnivores take advantage of periods of less human disturbance to satisfy their feeding needs. In addition, although the most frequent species in our review were a medium size (Figure 2), the selected models did not include the body mass variable, which may be attributed to our relatively small sample size.

Regarding taxonomic features, most of the species recorded in the articles are mammals and birds (mainly raptors), which may be explained by the general bias present in urban ecology towards these groups [60]. In the case of predators and scavengers, there were not many options beyond them; however, we still found some articles focused on Asian monitor lizards [96,97], and others that include these reptiles as scavengers among other species [94]. Despite that, the selected models did not include any taxonomic classification, but the model that included the class variable proved to be better than the null model, being very close to the threshold of 2 points of AICc. This, added to the fact that the presence among the main species detected in both cases includes mammals such as the red fox, and the domestic dog and cat, makes us suspect that there may indeed be a taxonomic tendency among carnivores to adapt to cities. However, available data in the reviewed articles were not sufficient to demonstrate this.

4.4. Conservation Implications

Regarding our results, we consider that it is necessary to devote more effort to assessing the role that cities may play for scavengers, in order to obtain a more complete understanding of the ecological and conservation implications which may result from the urban predatory and scavenging habits of wildlife [11,33]. In this sense, from a conservation point of view, species management strategies should consider the role of cities for predators and scavengers, due to the short- and long-term influence (both positive and negative) that their presence in cities and the consumption of living prey and refuse may have, at both individual and population levels [28–30,45]. Due to natural habitat loss, the probability of larger mammals colonizing urban environments may increase, as evidenced by the anecdotal but growing data shown for leopards (*Panthera pardus*), tigers (*Panthera tigris*) and lions (*Panthera leo persica*) in India [98–100]. In this sense, the increase in the number of medium- to large-size predators feeding in cities may lead to an increase of potential human–wildlife conflicts, which will ultimately call for a better understanding of their presence in humanized areas. Furthermore, cities should be treated as other habitats in conservation projects relating to endangered carnivores, taking into account that they undoubtedly use this type of ecosystem. Thus, ignoring the pros and cons that result from the species' exploitation of urban environments would be negligent.

Lastly, we must reflect on the multiple interpretations from the nature conservationist point of view that the prominent presence of cats and dogs in both reviews offers. First, the predominance of the same species among the most recorded predators and scavengers is a logical consequence of their close association to humans, who have aided in their expansion wherever they have settled [101,102]. Beyond their presence, growing research shows how these species have a negative impact on native wildlife, thereby converting themselves into conflictive invasive species [103,104]. In the case of cities, the presence of cats and dogs (domestic or feral) may increase spatial-use conflicts with wildlife in urban green areas [105,106], also affecting native mammal, bird, reptile and amphibian communities by predation [107]. Finally, the high densities of these species can threaten public health by increasing the prevalence of diseases such as toxoplasma and rabies [108,109].

5. Conclusions

The expanding trend of urbanization, especially in developing countries [110,111] will likely lead to an increasing number of carnivores interacting in cities, due to the high availability of prey, garbage and organic remains, especially in countries with poor waste management practices. We also consider that, as the number of studies increases, more synanthropic species of medium and small size will be reported among the lists of carnivores exploiting cities, especially for those cities and countries for which we do not yet have scientific studies on urban predators and scavengers. Beyond birds and mammals, a priori, there are few options from other animal groups scavenging in cities, with the exception of some large reptiles. As such, the occurrence of different large monitor lizards (mainly *Varanus salvator* and *Varanus benghalensis*) in Asian cities may be an interesting topic to study in the coming years, attending to the role that they play as scavengers, and the potential conflict generated due to their presence. In general, we urge researchers to continue studying urban carnivores' diet, expanding their research to more species and regions with less scientific information. In addition, from the point of view of urban ecology, a greater knowledge of predators and scavengers would help to improve our understanding of the relationship between carnivores and their environment, in the case of human-dominated areas such as cities. Thus, future studies on predators and scavengers in cities should not only focus on the type of food urban carnivores consume but also on the location and distribution of meat resources within urban areas, their availability and predictability, and how these features influence the ecology of these species. Finally, a more complete comprehension of the presence of carnivores in urban areas may help to better inform citizens and mitigate conflicts with residents, improving the perception and acceptance of carnivores in urbanized areas, and, by extension, fostering an improved attitude toward nature in a broader sense.

Supplementary Materials: The following are available online at https://www.mdpi.com/1424-2818/13/2/46/s1, Table S1: Articles recorded during the urban predation search, Table S2: Articles recorded during the urban scavenging search, Table S3: Articles recorded in the urban predation review, Table S4: Articles recorded in the urban scavenging.

Author Contributions: Conceptualization, Á.L., P.R.-V. and E.A.; methodology, Á.L., P.R.-V. and E.A.; formal analysis E.A.; investigation, Á.L., P.R.-V. and E.A.; data curation, Á.L. and E.A.; writing—original draft preparation, Á.L.; writing—review and editing, Á.L., P.R.-V. and E.A.; visualization, Á.L., P.R.-V. and E.A.; supervision, Á.L., P.R.-V. and E.A. All authors have read and agreed to the published version of the manuscript.

Funding: The research was funded by RTI2018-099609-B-C21 (Spanish Ministry of Economy and Competitiveness and EU/ERDF).

Institutional Review Board Statement: Not applicable.

Informed Consent Statement: Not applicable.

Data Availability Statement: The data presented in this study are available in supplementary material.

Acknowledgments: We thank Ana Benítez (Doñana Biological Station) for her help and useful suggestions. The four anonymous reviewer made very valuable comments that greatly improved the manuscript, and Corrine O'Sullivan and Wouter Marc Gerard Vansteelant revised the English.

Conflicts of Interest: The authors declare no conflict of interest. The funders had no role in the design of the study; in the collection, analyses, or interpretation of data; in the writing of the manuscript, or in the decision to publish the results.

References

1. Dearborn, D.C.; Kark, S. Motivations for conserving urban biodiversity. *Conserv. Biol.* **2010**, *24*, 432–440. [CrossRef]
2. Aronson, M.F.; La Sorte, F.A.; Nilon, C.H.; Katti, M.; Goddard, M.A.; Lepczyk, C.A.; Warren, P.S.; Williams, N.S.; Cilliers, S.; Clarkson, B.; et al. A global analysis of the impacts of urbanization on bird and plant diversity reveals key anthropogenic drivers. *Proc. R. Soc. B* **2014**, *281*, 20133330. [CrossRef]
3. Braaker, S.; Ghazoul, J.; Obrist, M.K.; Moretti, M. Habitat connectivity shapes urban arthropod communities: The key role of green roofs. *Ecology* **2014**, *95*, 1010–1021. [CrossRef]
4. Threlfall, C.G.; Walker, K.; Williams, N.S.; Hahs, A.K.; Mata, L.; Stork, N.; Livesley, S.J. The conservation value of urban green space habitats for Australian native bee communities. *Biol. Conserv.* **2015**, *187*, 240–248. [CrossRef]
5. Clergeau, P.; Croci, S.; Jokimäki, J.; Kaisanlahti-Jokimäki, M.L.; Dinetti, M. Avifauna homogenisation by urbanisation: Analysis at different European latitudes. *Biol. Conserv.* **2006**, *127*, 336–344. [CrossRef]
6. Ortega-Álvarez, R.; MacGregor-Fors, I. Living in the big city: Effects of urban land-use on bird community structure, diversity, and composition. *Landsc. Urban Plan.* **2009**, *90*, 189–195. [CrossRef]
7. Faeth, S.H.; Bang, C.; Saari, S. Urban biodiversity: Patterns and mechanisms. *Ann. N. Y. Acad. Sci.* **2011**, *1223*, 69–81. [CrossRef]
8. Rebolo-Ifrán, N.; Tella, J.L.; Carrete, M. Urban conservation hotspots: Predation release allows the grassland-specialist burrowing owl to perform better in the city. *Sci. Rep.* **2017**, *7*, 1–9. [CrossRef]
9. Recio, M.R.; Payne, K.; Seddon, P.J. Emblematic forest dwellers reintroduced into cities: Resource selection by translocated juvenile kaka. *Curr. Zool.* **2016**, *62*, 15–22. [CrossRef]
10. Luna, Á.; Romero-Vidal, P.; Hiraldo, F.; Tella, J.L. Cities may save some threatened species but not their ecological functions. *PeerJ* **2018**, *6*, e4908. [CrossRef]
11. Soanes, K.; Lentini, P.E. When cities are the last chance for saving species. *Front. Ecol. Environ.* **2019**, *17*, 225–231. [CrossRef]
12. Thomas, J.P.; Jung, T.S. Life in a northern town: Rural villages in the boreal forest are islands of habitat for an endangered bat. *Ecosphere* **2019**, *10*, e02563. [CrossRef]
13. Kinzig, A.P.; Warren, P.; Martin, C.; Hope, D.; Katti, M. The effects of human socioeconomic status and cultural characteristics on urban patterns of biodiversity. *Ecol. Soc.* **2005**, *10*, 1–13. [CrossRef]

14. Lepczyk, C.A.; Aronson, M.F.; Evans, K.L.; Goddard, M.A.; Lerman, S.B.; MacIvor, J.S. Biodiversity in the city: Fundamental questions for understanding the ecology of urban green spaces for biodiversity conservation. *BioScience* **2017**, *67*, 799–807. [CrossRef]

15. Cardoso, G.C.; Klingbeil, B.T.; La Sorte, F.A.; Lepczyk, C.A.; Fink, D.; Flather, C.H. Exposure to noise pollution across North American passerines supports the noise filter hypothesis. *Glob. Ecol. Biogeogr.* **2020**, *29*, 1430–1434. [CrossRef]

16. Gaston, K.J.; Visser, M.E.; Hölker, F. The biological impacts of artificial light at night: The research challenge. *Philos. Trans. R. Soc. B* **2015**, *370*, 20140133. [CrossRef]

17. Ciach, M.; Fröhlich, A. Habitat preferences of the Syrian Woodpecker *Dendrocopos syriacus* in urban environments: An ambiguous effect of pollution. *Bird Study* **2013**, *60*, 491–499. [CrossRef]

18. Foltz, S.L.; Ross, A.E.; Laing, B.T.; Rock, R.P.; Battle, K.E.; Moore, I.T. Get off my lawn: Increased aggression in urban song sparrows is related to resource availability. *Behav. Ecol.* **2015**, *26*, 1548–1557. [CrossRef]

19. Kark, S.; Iwaniuk, A.; Schalimtzek, A.; Banker, E. Living in the city: Can anyone become an 'urban exploiter'? *J. Biogeogr.* **2007**, *34*, 638–651. [CrossRef]

20. Møller, A.P. Successful city dwellers: A comparative study of the ecological characteristics of urban birds in the Western Palearctic. *Oecologia* **2009**, *159*, 849–858. [CrossRef]

21. Evans, K.L.; Chamberlain, D.E.; Hatchwell, B.J.; Gregory, R.D.; Gaston, K.J. What makes an urban bird? *Glob. Chang. Biol.* **2011**, *7*, 32–44. [CrossRef]

22. Leveau, L.M. Bird traits in urban–rural gradients: How many functional groups are there? *J. Ornithol.* **2013**, *154*, 655–662. [CrossRef]

23. Baker, P.J.; Ansell, R.J.; Dodds, P.A.A.; Webber, C.E.; Harris, S. Factors affecting the distribution of small mammals in an urban area. *Mammal Rev.* **2003**, *33*, 95–100. [CrossRef]

24. Santini, L.; González-Suárez, M.; Russo, D.; Gonzalez-Voyer, A.; von Hardenberg, A.; Ancillotto, L. One strategy does not fit all: Determinants of urban adaptation in mammals. *Ecol. Lett.* **2019**, *22*, 365–376. [CrossRef]

25. Lowry, H.; Lill, A.; Wong, B.B. Behavioural responses of wildlife to urban environments. *Biol. Rev.* **2013**, *88*, 537–549. [CrossRef]

26. Murray, M.; Cembrowski, A.; Latham, A.D.M.; Lukasik, V.M.; Pruss, S.; St Clair, C.C. Greater consumption of protein-poor anthropogenic food by urban relative to rural coyotes increases diet breadth and potential for human–wildlife conflict. *Ecography* **2015**, *38*, 1235–1242. [CrossRef]

27. de Araujo, G.M.; Peres, C.A.; Baccaro, F.B.; Guerta, R.S. Urban waste disposal explains the distribution of Black Vultures (*Coragyps atratus*) in an Amazonian metropolis: Management implications for birdstrikes and urban planning. *PeerJ* **2018**, *6*, e5491. [CrossRef]

28. Houston, D.C.; Mee, A.; McGrady, M. Why do condors and vultures eat junk?: The implications for conservation. *J. Raptor Res.* **2007**, *41*, 235–238. [CrossRef]

29. Oro, D.; Genovart, M.; Tavecchia, G.; Fowler, M.S.; Martínez-Abraín, A. Ecological and evolutionary implications of food subsidies from humans. *Ecol. Lett.* **2013**, *16*, 1501–1514. [CrossRef]

30. Plaza, P.I.; Lambertucci, S.A. How are garbage dumps impacting vertebrate demography, health, and conservation? *Glob. Ecol. Conserv.* **2017**, *12*, 9–20. [CrossRef]

31. Hindmarch, S.; Elliott, J.E. When owls go to town: The diet of urban Barred Owls. *J. Raptor. Res.* **2015**, *49*, 66–74. [CrossRef]

32. McPherson, S.C.; Brown, M.; Downs, C.T. Diet of the crowned eagle (*Stephanoaetus coronatus*) in an urban landscape: Potential for human-wildlife conflict? *Urban Ecosyst.* **2016**, *19*, 383–396. [CrossRef]

33. Bateman, P.W.; Fleming, P.A. Big city life: Carnivores in urban environments. *J. Zool.* **2012**, *287*, 1–23. [CrossRef]

34. Newsome, S.D.; Garbe, H.M.; Wilson, E.C.; Gehrt, S.D. Individual variation in anthropogenic resource use in an urban carnivore. *Oecologia* **2015**, *178*, 115–128. [CrossRef] [PubMed]

35. Bildstein, K.L.; Therrien, J.F. Urban birds of prey: A lengthy history of human-raptor cohabitation. In *Urban Raptors*; Island Press: Washington, DC, USA, 2018; pp. 3–17.

36. Contesse, P.; Hegglin, D.; Gloor, S.; Bontadina, F.; Deplazes, P. The diet of urban foxes (*Vulpes vulpes*) and the availability of anthropogenic food in the city of Zurich, Switzerland. *Mamm. Biol.* **2004**, *69*, 81–95. [CrossRef]

37. Kristan, W.B., III; Boarman, W.I.; Crayon, J.J. Diet composition of common ravens across the urban-wildland interface of the West Mojave Desert. *Wildl. Soc. Bull.* **2004**, *32*, 244–253. [CrossRef]

38. Allen, B.L.; Carmelito, E.; Amos, M.; Goullet, M.S.; Allen, L.R.; Speed, J.; Leung, L.K.P. Diet of dingoes and other wild dogs in peri-urban areas of north-eastern Australia. *Sci. Rep.* **2016**, *6*, 1–8. [CrossRef]

39. Morgan, S.A.; Hansen, C.M.; Ross, J.G.; Hickling, G.J.; Ogilvie, S.C.; Paterson, A.M. Urban cat (*Felis catus*) movement and predation activity associated with a wetland reserve in New Zealand. *Wild. Res.* **2009**, *36*, 574–580. [CrossRef]

40. Henriques, M.; Granadeiro, J.P.; Hamilton Monteiro, A.N.; Lecoq, M.; Cardoso, P.; Regalla, A.; Catry, P. Not in wilderness: African vulture strongholds remain in areas with high human density. *PLoS ONE* **2018**, *13*, e0190594. [CrossRef]

41. Larson, R.N.; Morin, D.J.; Wierzbowska, I.A.; Crooks, K.R. Food habits of coyotes, gray foxes, and bobcats in a coastal southern California urban landscape. *West. N. Am. Nat.* **2015**, *75*, 339–347. [CrossRef]

42. Méndez, A.; Montalvo, T.; Aymí, R.; Carmona, M.; Figuerola, J.; Navarro, J. Adapting to urban ecosystems: Unravelling the foraging ecology of an opportunistic predator living in cities. *Urban Ecosyst.* **2020**, *23*, 1117–1126. [CrossRef]

43. Sims, V.; Evans, K.L.; Newson, S.E.; Tratalos, J.A.; Gaston, K.J. Avian assemblage structure and domestic cat densities in urban environments. *Divers. Distrib.* **2008**, *14*, 387–399. [CrossRef]
44. Drewitt, E.J.; Dixon, N. Diet and prey selection of urban-dwelling Peregrine Falcons in southwest England. *Br. Birds* **2008**, *101*, 58.
45. Braczkowski, A.R.; O'Bryan, C.J.; Stringer, M.J.; Watson, J.E.; Possingham, H.P.; Beyer, H.L. Leopards provide public health benefits in Mumbai, India. *Front. Ecol. Environ.* **2018**, *16*, 176–182. [CrossRef]
46. Shochat, E.; Warren, P.S.; Faeth, S.H.; McIntyre, N.E.; Hope, D. From patterns to emerging processes in mechanistic urban ecology. *Trends Ecol. Evol.* **2006**, *21*, 186–191. [CrossRef]
47. Rodewald, A.D.; Kearns, L.J.; Shustack, D.P. Anthropogenic resource subsidies decouple predator–prey relationships. *Ecol. App.* **2011**, *21*, 936–943. [CrossRef]
48. Fischer, J.D.; Cleeton, S.H.; Lyons, T.P.; Miller, J.R. Urbanization and the predation paradox: The role of trophic dynamics in structuring vertebrate communities. *Bioscience* **2012**, *62*, 809–818. [CrossRef]
49. Inger, R.; Cox, D.T.; Per, E.; Norton, B.A.; Gaston, K.J. Ecological role of vertebrate scavengers in urban ecosystems in the UK. *Ecol. Evol.* **2016**, *6*, 7015–7023. [CrossRef]
50. Schwartz, A.L.; Williams, H.F.; Chadwick, E.; Thomas, R.J.; Perkins, S.E. Roadkill scavenging behaviour in an urban environment. *J. Urban Ecol.* **2018**, *4*, juy006. [CrossRef]
51. Plaza, P.I.; Lambertucci, S.A. More massive but potentially less healthy: Black vultures feeding in rubbish dumps differed in clinical and biochemical parameters with wild feeding birds. *PeerJ* **2018**, *6*, e4645. [CrossRef]
52. Lambertucci, S.A.; Speziale, K.L.; Rogers, T.E.; Morales, J.M. How do roads affect the habitat use of an assemblage of scavenging raptors? *Biodivers. Conserv.* **2009**, *18*, 2063–2074. [CrossRef]
53. Arrondo, E.; Sanz-Aguilar, A.; Pérez-García, J.M.; Cortés-Avizanda, A.; Sánchez-Zapata, J.A.; Donázar, J.A. Landscape anthropization shapes the survival of a top avian scavenger. *Biodivers. Conserv.* **2020**, *29*, 1411–1425. [CrossRef]
54. Ogada, D.L.; Keesing, F.; Virani, M.Z. Dropping dead: Causes and consequences of vulture population declines worldwide. *Ann. N. Y. Acad. Sci.* **2012**, *1249*, 57–71. [CrossRef] [PubMed]
55. Buechley, E.R.; McGrady, M.J.; Çoban, E.; Şekercioğlu, Ç.H. Satellite tracking a wide-ranging endangered vulture species to target conservation actions in the Middle East and East Africa. *Biodivers. Conserv.* **2018**, *27*, 2293–2310. [CrossRef]
56. Novaes, W.G.; Abreu, T.L.D.S.; Guerta, R.S. Assessing vulture translocation as a management tool to mitigate airport bird strikes. *Hum.-Wildl. Interact.* **2020**, *14*, 19.
57. Parra, J.; Tellería, J.L. The increase in the Spanish population of Griffon Vulture *Gyps fulvus* during 1989–1999: Effects of food and nest site availability. *Bird Conserv. Int.* **2004**, *14*, 33–41. [CrossRef]
58. Moleón, M.; Sanchez-Zapata, J.A.; Margalida, A.; Carrete, M.; Owen-Smith, N.; Donazar, J.A. Humans and scavengers: The evolution of interactions and ecosystem services. *BioScience* **2014**, *64*, 394–403. [CrossRef]
59. McDonnell, M.J.; Niemelä, J. The history of urban ecology. *Urban Ecol.* **2011**, *9*, 34–49.
60. Magle, S.B.; Hunt, V.M.; Vernon, M.; Crooks, K.R. Urban wildlife research: Past, present, and future. *Biol. Conserv.* **2012**, *155*, 23–32. [CrossRef]
61. McPhearson, T.; Pickett, S.T.; Grimm, N.B.; Niemelä, J.; Alberti, M.; Elmqvist, T.; Qureshi, S. Advancing urban ecology toward a science of cities. *BioScience* **2016**, *66*, 198–212. [CrossRef]
62. DeVault, T.L.; Rhodes, O.E., Jr.; Shivik, J.A. Scavenging by vertebrates: Behavioral, ecological, and evolutionary perspectives on an important energy transfer pathway in terrestrial ecosystems. *Oikos* **2003**, *102*, 225–234.
63. Cincotta, R.P.; Wisnewski, J.; Engelman, R. Human population in the biodiversity hotspots. *Nature* **2000**, *404*, 990–992. [CrossRef] [PubMed]
64. Bradshaw, C.J.; Sodhi, N.S.; Brook, B.W. Tropical turmoil: A biodiversity tragedy in progress. *Front. Ecol. Environ.* **2009**, *7*, 79–87. [CrossRef]
65. Haddaway, N.R.; Woodcock, P.; Macura, B.; Collins, A. Making literature reviews more reliable through application of lessons from systematic reviews. *Conserv. Biol.* **2015**, *29*, 1596–1605. [CrossRef] [PubMed]
66. Lozano, J.; Olszańska, A.; Morales-Reyes, Z.; Castro, A.A.; Malo, A.F.; Moleón, M.; Sánchez-Zapata, J.A.; Cortés-Avizanda, A.; von Wehrden, H.; Dorresteijn, I.; et al. Human-carnivore relations: A systematic review. *Biol. Conserv.* **2019**, *237*, 480–492. [CrossRef]
67. Haddaway, N.R.; Bayliss, H.R. Shades of grey: Two forms of grey literature important for reviews in conservation. *Biol. Conserv.* **2015**, *191*, 827–829. [CrossRef]
68. Paez, A. Gray literature: An important resource in systematic reviews. *J. Evid. Based. Med.* **2017**, *10*, 233–240. [CrossRef]
69. Olson, D.M.; Dinerstein, E.; Wikramanayake, E.D.; Burgess, N.D.; Powell, G.V.N.; Underwood, E.C.; D'amico, J.A.; Itoua, I.; Strand, H.E.; Morrison, J.C.; et al. Terrestrial Ecoregions of the World: A New Map of Life on Earth: A new global map of terrestrial ecoregions provides an innovative tool for conserving biodiversity. *BioScience* **2001**, *51*, 933–938. [CrossRef]
70. Wilman, H.; Belmaker, J.; Simpson, J.; de la Rosa, C.; Rivadeneira, M.M.; Jetz, W. EltonTraits 1.0: Species-level foraging attributes of the world's birds and mammals: Ecological Archives E095-178. *Ecology* **2014**, *95*, 2027. [CrossRef]
71. Shine, R.; Harlow, P.S. Ecological traits of commercially harvested water monitors, *Varanus salvator*, in northern Sumatra. *Wildl. Res.* **1998**, *25*, 437–447. [CrossRef]
72. Guarino, F. Diet of a large carnivorous lizard, *Varanus varius*. *Wildl. Res.* **2001**, *28*, 627–630. [CrossRef]
73. Guarino, F. Spatial ecology of a large carnivorous lizard, *Varanus varius* (Squamata: Varanidae). *J. Zool.* **2002**, *258*, 449–457. [CrossRef]

74. Martins, F.I.; De Souza, F.L.; Da Costa, H.T.M. Feeding habits of *Phrynops geoffroanus* (Chelidae) in an urban river in Central Brazil. *Chelonian Conserv. Biol.* **2010**, *9*, 294–297. [CrossRef]
75. Balakrishna, S.; Batabyal, A.; Thaker, M. Dining in the city: Dietary shifts in Indian rock agamas across an urban–rural landscape. *J. Herpetol.* **2016**, *50*, 423–428. [CrossRef]
76. Burnham, K.P.; Anderson, D.R. *Model Selection and Multimodel Inference: A Practical Information-Theoretic Approach (2ª), Ecological Modelling*; Springer Science & Business Media: Berlin/Heidelberg, Germany, 2002.
77. Arnold, T.W. Uninformative parameters and model selection using Akaike's Information Criterion. *J. Wildl. Manag.* **2010**, *74*, 1175–1178. [CrossRef]
78. Kleiber, C.; Zeileis, A. *Applied Econometrics with R*; Springer: New York, NY, USA, 2008; ISBN 978-0-387-77316-2.
79. R Development Core Team. *R: A Language and Environment for Statistical Computing (R-3.6.1)*; R Foundation for StatisticalComputing: Vienna, Austria, 2019.
80. Sazima, I. Australian Raven (*Corvus coronoides*) scavenges on all five major vertebrate groups at urban Sydney, Southeast Australia. *Trop. Nat. Hist.* **2020**, *20*, 89–94.
81. Clifton, B.; Jones, D.N. Finding food in a human-dominated environment: Exploring the foraging behaviour of urban Torresian Crows' Corvus orru'. *Aust. Field Ornithol.* **2017**, *34*, 30. [CrossRef]
82. Schlacher, T.A.; Weston, M.A.; Lynn, D.; Schoeman, D.S.; Huijbers, C.M.; Olds, A.D.; Connolly, R.M. Conservation gone to the dogs: When canids rule the beach in small coastal reserves. *Biodivers. Conserv.* **2015**, *24*, 493–509. [CrossRef]
83. Welti, N.; Scherler, P.; Grüebler, M.U. Carcass predictability but not domestic pet introduction affects functional response of scavenger assemblage in urbanized habitats. *Funct. Ecol.* **2020**, *34*, 265–275. [CrossRef]
84. Wu, J. Urban ecology and sustainability: The state-of-the-science and future directions. *Landsc. Urban Plan.* **2014**, *125*, 209–221. [CrossRef]
85. Beasley, J.C.; Olson, Z.H.; DeVault, T.L. Ecological role of vertebrate scavengers. In *Carrion Ecology, Evolution and Their Applications*, 1st ed.; Eric, B., Jeffery, K.T., Aaron, M.T., Eds.; CRC Press: Cleveland, OH, USA, 2015; pp. 107–127.
86. King, D. The scientific impact of nations. *Nature* **2004**, *430*, 311–316. [CrossRef]
87. Hoornweg, D.; Bhada-Tata, P. What a waste: A global review of solid waste management. Urban development series; knowledge papers no. 15. World Bank: Washington, DC, USA, 2012.
88. Mudzengerere, F.H.; Chigwenya, A. Waste Management in Bulawayo city council in Zimbabwe: In search of Sustainable waste Management in the city. *J. Sustain. Dev. Afr.* **2012**, *14*, 228–244.
89. Abay, G.Y.; Bauer, H.; Gebrihiwot, K.; Deckers, J. Peri-urban spotted hyena (*Crocuta crocuta*) in northern Ethiopia: Diet, economic impact, and abundance. *Eur. J. Wildl. Res.* **2011**, *57*, 759–765. [CrossRef]
90. Sol, D.; Gonzalez-Lagos, C.; Moreira, D.; Maspons, J. Urbanisation tolerance and the loss of avian diversity. *Ecol. Lett.* **2014**, *17*, 942–950. [CrossRef]
91. Carrete, M.; Tella, J.L. Behavioral correlations associated with fear of humans differ between rural and urban burrowing owls. *Front. Ecol. Evol.* **2017**, *5*, 54. [CrossRef]
92. Beninde, J.; Veith, M.; Hochkirch, A. Biodiversity in cities needs space: A meta-analysis of factors determining intra-urban biodiversity variation. *Ecol. Lett.* **2015**, *18*, 581–592. [CrossRef] [PubMed]
93. Iserhard, C.A.; Duarte, L.; Seraphim, N.; Freitas, A.V.L. How urbanization affects multiple dimensions of biodiversity in tropical butterfly assemblages. *Biodivers. Conserv.* **2019**, *28*, 621–638. [CrossRef]
94. Huijbers, C.M.; Schlacher, T.A.; Schoeman, D.S.; Weston, M.A.; Connolly, R.M. Urbanisation alters processing of marine carrion on sandy beaches. *Landsc Urban Plan.* **2013**, *119*, 1–8. [CrossRef]
95. Riding, C.S.; Loss, S.R. Factors influencing experimental estimation of scavenger removal and observer detection in bird–window collision surveys. *Ecol. Appl.* **2018**, *28*, 2119–2129. [CrossRef]
96. Kulabtong, S.; Mahaprom, R. Observation on food items of Asian water monitor, *Varanus salvator* (Laurenti, 1768) (Squamata Varanidae), in urban eco-system, Central Thailand. *Biodivers. J.* **2014**, *6*, 695–698.
97. Rahman, K.M.; Khan, M.M.H.; Rakhimov, I.I. Scavenging Behavior of the Bengal Monitor (Varanus bengalensis) in Jahangirnagar University Campus, Bangladesh. *J. Sci. Res. Rep.* **2015**, 539–550. [CrossRef]
98. Karanth, K.U.; Gopal, R. An ecology-based policy framework for human-tiger coexistence in India. *Conserv. Biol. Ser.-Camb.* **2005**, *9*, 373–387.
99. Meena, V.; Jhala, Y.V.; Chellam, R.; Pathak, B. Implications of diet composition of Asiatic lions for their conservation. *J. Zool.* **2011**, *284*, 60–67. [CrossRef]
100. Kumbhojkar, S.; Yosef, R.; Kosicki, J.Z.; Kwiatkowska, P.K.; Tryjanowski, P. Dependence of the leopard Panthera pardus fusca in Jaipur, India, on domestic animals. *Oryx* **2020**, 1–7. [CrossRef]
101. Medina, F.M.; Bonnaud, E.; Vidal, E.; Tershy, B.R.; Zavaleta, E.S.; Josh Donlan, C.; Nogales, M. A global review of the impacts of invasive cats on island endangered vertebrates. *Glob. Chang. Biol.* **2011**, *17*, 3503–3510. [CrossRef]
102. Young, J.K.; Olson, K.A.; Reading, R.P.; Amgalanbaatar, S.; Berger, J. Is wildlife going to the dogs? Impacts of feral and free-roaming dogs on wildlife populations. *Bioscience* **2011**, *61*, 125–132. [CrossRef]
103. Loss, S.R.; Will, T.; Marra, P.P. The impact of free-ranging domestic cats on wildlife of the United States. *Nat. Commun.* **2013**, *4*, 1–8. [CrossRef]

104. Hughes, J.; Macdonald, D.W. A review of the interactions between free-roaming domestic dogs and wildlife. *Biol. Conserv.* **2013**, *157*, 341–351. [CrossRef]
105. Coronel-Arellano, H.; Rocha-Ortega, M.; Gual-Sill, F.; Martínez-Meyer, E.; Ramos-Rendón, A.K.; González-Negrete, M.; Zambrano, L. Raining feral cats and dogs? Implications for the conservation of medium-sized wild mammals in an urban protected area. *Urban Ecosyst.* **2020**, 1–12. [CrossRef]
106. Mella-Méndez, I.; Flores-Peredo, R.; Bolívar-Cimé, B.; Vázquez-Domínguez, G. Effect of free-ranging dogs and cats on medium-sized wild mammal assemblages in urban protected areas of a Mexican city. *Wildl. Res.* **2020**, *46*, 669–678. [CrossRef]
107. Seymour, C.L.; Simmons, R.E.; Morling, F.; George, S.T.; Peters, K.; O'Riain, M.J. Caught on camera: The impacts of urban domestic cats on wild prey in an African city and neighbouring protected areas. *Glob. Ecol. Conserv.* **2020**, *23*, e01198. [CrossRef]
108. Markandya, A.; Taylor, T.; Longo, A.; Murty, M.N.; Murty, S.; Dhavala, K. Counting the cost of vulture decline—an appraisal of the human health and other benefits of vultures in India. *Ecol. Econ.* **2008**, *67*, 194–204. [CrossRef]
109. Lepczyk, C.A.; Haman, K.H.; Sizemore, G.C.; Farmer, C. Quantifying the presence of feral cat colonies and Toxoplasma gondii in relation to bird conservation areas on O'ahu, Hawai'i. *Conserv. Sci. Pract.* **2020**, *2*, e179. [CrossRef]
110. Seto, K.C.; Güneralp, B.; Hutyra, L.R. Global forecasts of urban expansion to 2030 and direct impacts on biodiversity and carbon pools. *Proc. Natl. Acad. Sci. USA* **2012**, *109*, 16083–16088. [CrossRef]
111. Chen, G.; Li, X.; Liu, X.; Chen, Y.; Liang, X.; Leng, J.; Xu, X.; Liao, W.; Qiu, Y.; Wu, Q.; et al. Global projections of future urban land expansion under shared socioeconomic pathways. *Nat. Commun.* **2020**, *11*, 1–12. [CrossRef] [PubMed]

Review

Visual Adaptations in Predatory and Scavenging Diurnal Raptors

Simon Potier

Department of Biology, Lund University, Sölvegatan 35, S-22362 Lund, Sweden; sim.potier@gmail.com

Received: 1 September 2020; Accepted: 14 October 2020; Published: 15 October 2020

Abstract: Ecological diversity among diurnal birds of prey, or raptors, is highlighted regarding their sensory abilities. While raptors are believed to forage primarily using sight, the sensory demands of scavengers and predators differ, as reflected in their visual systems. Here, I have reviewed the visual specialisations of predatory and scavenging diurnal raptors, focusing on (1) the anatomy of the eye and (2) the use of vision in foraging. Predators have larger eyes than scavengers relative to their body mass, potentially highlighting the higher importance of vision in these species. Scavengers possess one centrally positioned fovea that allows for the detection of carrion at a distance. In addition to the central fovea, predators have a second, temporally positioned fovea that views the frontal visual field, possibly for prey capture. Spatial resolution does not differ between predators and scavengers. In contrast, the organisation of the visual fields reflects important divergences, with enhanced binocularity in predators opposed to an enlarged field of view in scavengers. Predators also have a larger blind spot above the head. The diversity of visual system specializations according to the foraging ecology displayed by these birds suggests a complex interplay between visual anatomy and ecology, often unrelatedly of phylogeny.

Keywords: birds of prey; foraging; predators; scavengers; vision

1. Introduction

Understanding how animals extract information from their surrounding environment is essential to appreciate how they complete their daily tasks such as finding food, avoiding predators, attracting mates, and navigating their environment [1]. Among the sensory organs found in the animal kingdom, eyes can provide instantaneous information about the environment like no other [2]. While birds use a wide range of cues (ranging from the Earth's magnetic field to sounds, odours, visual cues, and so on), vision appears to be a very important sensory modality in these animals, especially in the context of foraging [3]. In particular, birds of prey, or "raptors" (as defined below), are considered to forage primary using the visual sense ([4,5], but see [6]). Indeed, a number of species of raptors spot their prey/food at relatively high altitudes where acoustic and olfactory cues should be undetectable, and vision may be the only accessible cue available for prey/food detection in the absence of visual barriers. This apparent reliance on vision is underlined by the fact that some raptor species have the highest visual acuity (also called spatial resolution) found to date, for both achromatic ([7,8], reviewed in [5]) and chromatic [9] patterns.

While precise terminology is essential in science, defining which species belong to "raptors" (or birds of prey) has been debated for decades [10]. Derived from the Latin word "rapere", "raptor" means plunderer or ravisher. Recently, McClure et al. (2019) defined raptors as "all species within orders that evolved from a raptorial landbird lineage and in which most species maintained their raptorial lifestyle as derived from their common ancestor" [10]. This definition, based on phylogeny, ecology, and morphology, includes species in the orders of Accipitriformes (hawks, eagles, Old World vultures, kites), Cathartiformes (New World vultures), Strigiformes (owls), Falconiformes (falcons and caracaras), and the unexpected Cariamiformes (seriemas) (see Figure 1). Because they include species that do not belong to a monophyletic clade [11] and with important morphological and ecological differences, raptors may differ significantly in their sensory abilities.

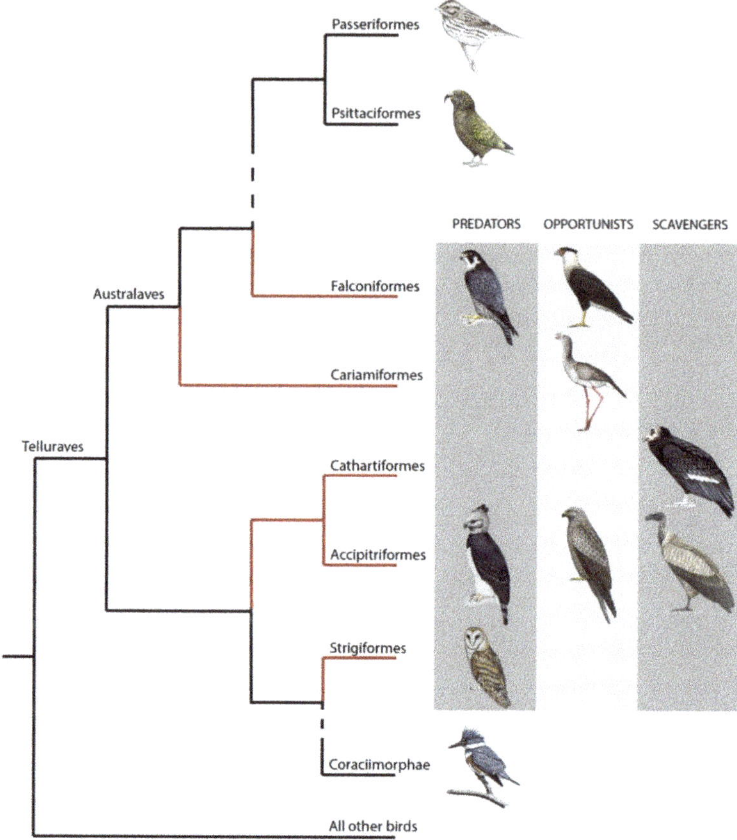

Figure 1. Phylogeny of core landbirds modified from [10]. The red branches encompass the order considered as raptors. For each order belonging to raptors, the presence/absence of the three foraging diet categories (predators, opportunists and scavengers) is represented by a drawing of a chosen species. "Raptor" is a paraphyletic group where species mostly share the raptorial lifestyle passed down from their single common ancestor [10]. This assumes that raptorial lifestyle has been lost twice independently with the ancestor of both Coraciimorphae and Passeriformes/Psittaciformes clades. Coraciimorphae contains six orders: Coliiformes (mousebirds), Trogoniformes (trogons), Coraciiformes (roller, kingfishers, and bee-eaters), Piciformes (woodpeckers), Leptosomiformes (cuckoo-rollers), and Bucerotiformes (hoopoes and hornbills). Drawings from Bryce W. Robinson.

A notable ecological difference among diurnal raptors is that some are active predators, whereas others are obligate scavengers. While a number of predators may scavenge (e.g., Steppe eagle *Aquila nipalensis*, Common buzzard *Buteo buteo*, and others [12]) and a small number of obligate scavengers occasionally predate (e.g., White-headed vulture *Trigonoceps occipitalis* [13]), those events are not preponderant. Consequently, scavengers and predators are likely to differ markedly in their sensory demands.

In this review, I present the differences and similarities in the visual systems of predatory and scavenging diurnal raptors, focusing exclusively on Accipitriformes, Cathartiformes, and Falconiformes. Interestingly, while the order Accipitriformes contains predators, opportunists and obligate scavengers, the order Cathartiformes only contains obligate scavenger species. Furthermore, the order Falconiformes only contains predators and some opportunist species (caracaras). Cariamiformes were not included in this review as very little (if anything) is known about their visual abilities. I also decided to not include owls (from Strigiformes) because of their nocturnal lifestyle, which implies different specialization/adaptation of the visual system [5]. Furthermore, even though some owls may scavenge on occasion [14–17], there are no obligate scavenging owl species, making the comparison between scavengers and predators impossible in these birds.

Previous reviews described the visual system of raptors [4,5], however, they did not concentrate on the differences and similarities among raptors with different foraging tactics. The aim of this review is to concentrate on visual adaptations of diurnal raptors with different lifestyle. Throughout the review, raptors were categorized as predators, scavengers, or opportunists according to Wilman et al. (2014) [18]. When species were not present in Wilman et al. (2014), foraging specialization was categorized according to Ferguson-Lees and Christie (2001) [12]. Specifically, all vultures were considered as scavengers. Caracaras and species that scavenge at least 40% of their time were considered as opportunists. Finally, species that scavenge occasionally (<20% of their time) were considered as predators. Furthermore, I also decided to highlight the lack of knowledge in the visual abilities of raptors, especially of scavengers. Emphasizing the sensory specializations of raptors with different foraging ecology and the little knowledge in scavenging species may have a crucial impact on the conservation programs of raptors.

2. Anatomical Specialization of the Eye

2.1. Predators Have Larger Eyes Than Scavengers Relative to Their Body Mass

In general, birds have big eyes in both relative (compared with body mass) and absolute terms [19], and raptors have relatively larger eyes than other birds [5,20]. Large eyes have long focal lengths and subsequently larger retinal images, and thus a potential for higher spatial resolving power in diurnal animals [2,21]. As a result, the relatively large eyes of raptors indicate the importance of the visual system for their daily life, especially because important costs are associated with increased eye size: (1) increased risk of being damaged, (2) mechanical and aerodynamic constraints [20], (3) higher metabolic and energetic costs [22], or (4) disability glare because of increased direct sunlight in the absence of adnexa (e.g., eye lids, eye brows [23]).

Interestingly, relative to their body mass, scavengers and opportunists have significantly smaller eyes than predators [24] (Figures 2A and 3). This would suggest that scavengers and opportunists invest less in vision and potentially might not need as high spatial resolution of vision as predatory species. However, from behavioural and anatomical studies, there is no evidence for lower spatial resolution in scavengers, except for Cathartiformes, whose eyes are also smaller than those of other groups (but not significantly when controlling for phylogeny [5]). Because the neural structures compete for space in the brain [25] and larger eyes may require a greater proportion of brain space dedicated to vision, scavengers and opportunists may invest more in other sensory modalities. For instance, Cathartiformes have larger olfactory bulbs than other raptors [26], and most species in which olfactory abilities have been shown are scavengers (see [6] for a review). However, olfactory bulbs appear to be freer to vary in size irrespective of other sensory structures [27] and very little is known about olfactory abilities in raptors in general [6]. More investigations are needed to understand why scavengers have smaller eyes than predators.

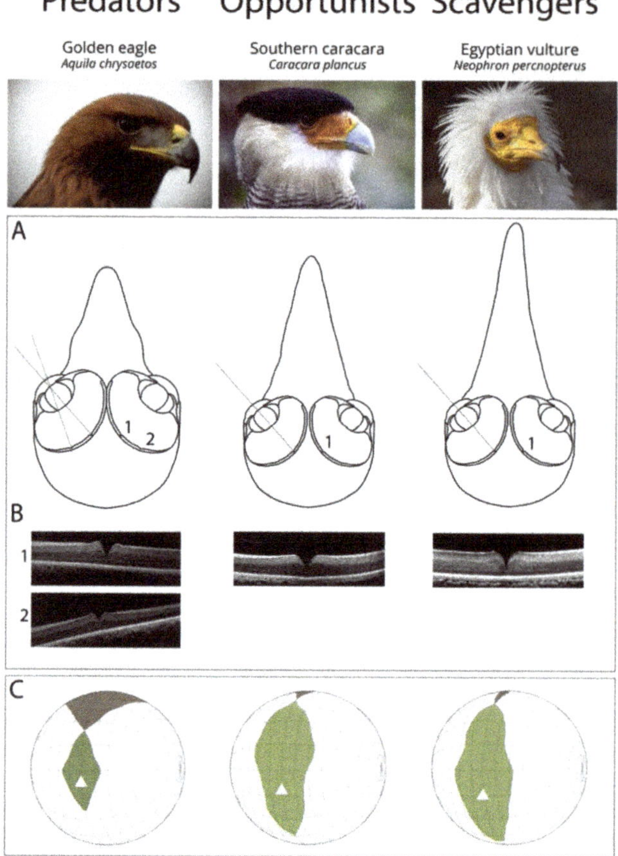

Figure 2. Functional differences of the visual system of raptors from different foraging tactics. (**A**) Schematic representation of frontal sections of the three chosen species (Golden eagle *Aquila chrysaetos*, Southern caracara *Caracara plancus*, and Egyptian vulture *Neophron percnopterus*) at the foveal plane. Fovea(s) and the centre of the pupil in each eye are on the plane. Grey lines represent the lines of sight of (1) the deep central fovea and (2) the shallow temporal fovea. Figures re-drawn from [28]. (**B**) Spectral domain optical coherence tomography (SD-OCT) images (B-scans) of the (1) central and (2) temporal fovea(s). Note that the Southern caracara and the Egyptian vulture lack temporal foveas. (**C**) Orthographic projection of retinal field boundaries of the eyes. A latitude and longitude coordinate system was used with the equator aligned vertically in the median sagittal plane (20 deg intervals in latitude and 10 deg intervals in longitude). The bird's head is at the centre of the globe. Green areas represent the binocular sector, white areas represent the monocular sectors, and brown areas represent the blind sectors. Triangles: direction of bill projection. Figures modified from [29,30]. Photography of the species was free of right thanks to @myb777_photography for the Golden eagle, @wal_172619 for the Southern caracara, and @pixel_mixer for the Egyptian vulture.

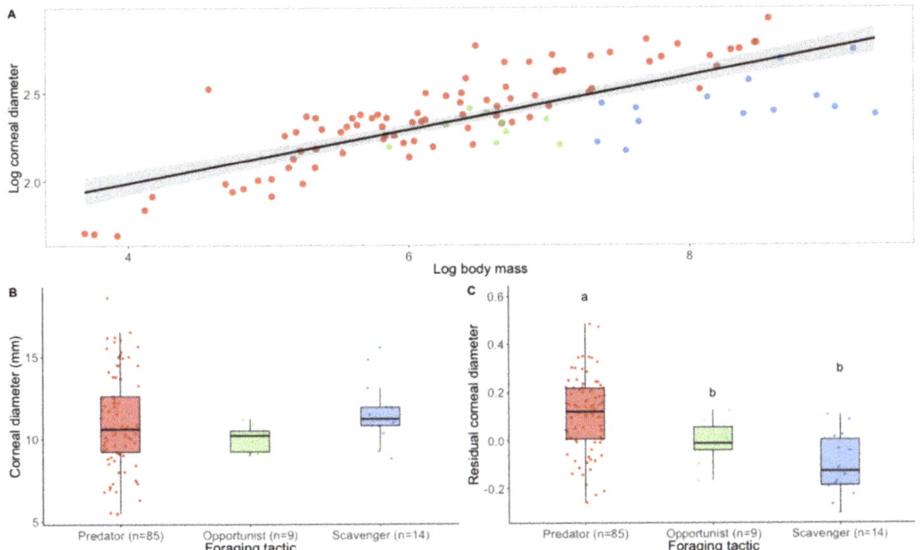

Figure 3. The eye size of raptors according to their foraging tactic. (**A**) Logarithmic relationship (black line) and 95% confidence level interval (grey shade) between corneal diameter (a proxy for eye size) and body mass in raptors (estimate = 0.14 ± 0.01, t = 9.48, p < 0.001). (**B**) Corneal diameter and (**C**) residual corneal diameter calculated from corneal diameter scaled to body weight in relation to foraging tactics. Differences among foraging tactic were tested using phylogenetic linear regression. The phylogenetic relationships among 130 species were estimated using a consensus tree based on 100 randomly selected trees from www.BirdTree.org [11] using Ericson tree distribution. Data were analysed on R 4.0.0 using ggplot2 [31], phylolm [32], phytools [33], caper [34], lmtest [35], ggpubr [36], and plyr [37]. Edge lengths were obtained by computing the mean edge length for each edge in the consensus tree. Model selection based on AICc and likelihood ratio test (lrtest function from lmtest package [35]) showed no differences for corneal diameter among foraging tactics (Chisq = 0.52, p = 0.77). By contrast, a significant difference was found for residual corneal diameter among foraging tactics (Chisq = 24.47, p < 0.001). Predators have significantly larger eyes compared with their body mass than opportunists (estimate = −0.11 ± 0.05, t = −2.20, p = 0.03) and scavengers (estimate = −0.21 ± 0.04, t = −4.90, p < 0.001). Scavengers and opportunists do not differ (estimate = −0.09 ± 0.06, t = −1.50, p = 0.14). Dots represent species. Different colours represent different foraging tactics (red = predator, green = opportunist, blue = scavenger). Different letters represent significant difference. Body mass were taken from [38] and corneal diameters from [27,39–41]. (Boxplots: black lines represent the median, coloured boxes represent the interquartile (IQR) range from 25th (Q1) to 75th (Q3) percentile, whiskers represent Q1 − 1.5 * IQR and Q3 + 1.5 * IQR). Note: for foraging tactics classification, please refer to Supplementary Table S1.

2.2. A Shared Optical System

All raptors, like all vertebrates, have camera-type eyes. Incoming light passes through the ocular media (cornea, aqueous humor, lens, and vitreous humor) and finally reaches the retina [2]. The eye can be described optically in a simple manner by the anterior focal length (which is correlated to axial length) and the pupil aperture [42]. Because pupil diameter sets the optical cut-off frequency for resolving power, the larger the pupil, the lower the diffraction limit and the higher the possible spatial resolution. Even in bright light conditions, raptors do not close their pupils [43], which again highlights the need for high spatial resolution in these species (but see [42] for optical aberrations).

The cornea and the lens function to focus the image on the retina. The accommodative power (in Diopter, i.e., the measure of the vergence of light corresponding as the reciprocal metre of the focal length) of the eye allows to maintain a clear image—or to focus on an object—as its distance varies. Therefore, a high range of accommodation is necessary for species that need to perform fine visual details at both close and long range and species that exploit different environments, such as cormorants that exploit both terrestrial and aquatic environments [44]. Birds can accommodate with the lens and the cornea [44]. Overall, across vertebrates, accommodative ability is related to lifestyle, with nocturnal species having lower accommodative power than diurnal animals [45]. The total accommodative power has been estimated in only one scavenger (Turkey vulture *Cathartes aura*: 8.5 Dioptres (D)) and six predators (Bald eagle *Haliaeetus leucocephalus*: 6.8 D; African fish eagle *Haliaeetus vocifer*: 9 D; Golden eagle *Aquila chrysaetos*: 6.7 D; American kestrel *Falco sparverius*: 16 D; Sharp-shinned hawk *Accipiter striatus*: 4 D; Red-tailed hawk *Buteo jamaicensis*: 25.8 D) in diurnal raptors [46]. It is still not clear why the Red-tailed hawk has significantly higher accommodative power than other diurnal raptors. Greater accommodation has only been measured in aquatic birds (e.g., 70–80D in waterfowls [47]). While other studies should be conducted, the accommodation capacity of the Turkey vulture falls in the range of predatory species. Glasser et al. (1997) suggested that all raptors need similar accommodative power in order to accurately position the beak for tearing at a carcass or a prey, and that this may be a stronger determinant of accommodative ability than catching prey with talons [46]. This hypothesis is supported by the similar accommodative power in seed-pecking birds (e.g., up to 17D in the chicken [48]), but the lower accommodation in owls (0–2D [46]), which swallow the whole prey and do not need accurate beak position.

2.3. Predators, but Not Scavengers, are Bifoveate

Raptors have the inverted retina design found in all vertebrates [2,24]. Raptors have a thicker central retina (400–500 µm thick [24]) than other birds (200–350 µm [49]). However, the retina is significantly thicker in predators [24] (Figure 2B), potentially highlighting their average higher peak retinal ganglion cell (RGC) densities [50–52]. Retinal ganglion cell density reflects the spatial resolving power of a species, and this technique has been used in Cathartiformes species [52]. We might expect predators to have higher spatial resolving powers than scavengers in view of their higher peak RGC (but see below). However, in species (such as raptors) with a fovea—an invagination in the inner retina where the photoreceptor density is the highest [53]—the RGC/cone ratio is 1:1 in the central fovea, indicating the cone density, not RGC density, limits spatial resolution [54,55].

The function of the physical structure of the fovea is still under debate [53,56], but interestingly, the number and position of fovea(s) vary in birds. All diurnal raptors studied so far possess at least one central fovea (a fovea that is centrally placed in the retina). The central fovea in raptors has been described as "convexiclivate" by Walls [57], who suggested that its steep slopes, together with the different refractive indices of the vitreous and retina, would serve as magnifying the image, and thus increase spatial resolution. Snyder and Miller (1978) suggested that this is only true for the bottom part of the foveal pit [58]. Predators and scavengers both display similar central foveal depth [24], which, according to Walls and Snyder and Miller's theories, would strongly indicate that both groups have similar magnification power. However, recent evidence suggests that fovea organisation in raptors is even more complex, as Potier et al. (2020) found that foveal shape varies significantly with age and eye size within one predatory species, the Black kite *Milvus migrans* [59].

Other authors have suggested that image magnification by the fovea is uneven [7,60] and should distort the image [61], except in the very bottom of the fovea, thus facilitating visual fixation. Interestingly, all predatory raptors studied to date (except the Broad-winged hawk *Buteo platypterus* [62]) possess a second fovea placed temporally in the retina, which is linked to frontal vision (Figure 2A,B) [24,51]. In contrast, scavengers lack this temporal fovea and, interestingly, opportunists differ in the presence or absence of the temporal fovea. Because the fovea should improve visual fixation [61], the temporal fovea has been suggested to be important for prey fixation at the moment of capture where frontal vision (Figure 2A,B) is necessary for accurate foot positioning [63]. As scavengers do not forage on highly manoeuvrable prey, they may not need a temporal fovea. This theory is supported by the ecological convergence between active predatory raptors and some non-raptorial species that pursue highly manoeuvrable prey, which also possess a temporal fovea, such as Least terns *Sternula antillarum* [50], Sacred kingfishers *Halycon sancta*, Laughing kookaburras *Dacelo novaeguineae* [64], and Tree swallow *Tachycineta bicolor* [65].

3. Foraging by Sight

3.1. The Spatial Resolution of Vision Does Not Depend on the Foraging Mode

Since Fox's work on falcon (American kestrel *Falco sparverius*) visual acuity, raptors have been considered to have a visual acuity at least three times that of humans (160 c/deg, i.e., 160 black and 160 white vertical bars in one degree of visual angle [66]). However, a few years later, Hirsch (1982) showed that the visual acuity of the American kestrel had been considerably overestimated by Fox et al. (1976), with a correct value closer to 40 c/deg [67].

To date, the visual acuity has been estimated in eleven diurnal raptor species, including four predators: Wedge tailed eagle *Aquila audax*, Harris's hawk *Parabuteo unicinctus*, American kestrel, and Brown falcon *Falco berigora*; five scavengers: Turkey vulture, Black vulture *Coragyps atratus*, Indian vulture *Gyps indicus*, Griffon vulture *Gyps fulvus*, and Egyptian vulture *Neophron percnopterus*; and two opportunists: Black kite *Milvus migrans* and Chimango caracara *Milvago chimango*. While different methods have been used (behaviour and anatomical estimation using either cone photoreceptor or RGC spacing), visual acuity has been found to vary from 15 c/deg in the Turkey vulture [52] to 142 c/deg in the Wedge tailed eagle [7]; see [4] for a review. A detailed table can be found in [5]. Interestingly, in behavioural studies on both for scavengers and predators, visual resolution drops rapidly as light levels fall [7,8,60].

Based on these 11 species, variation in visual acuity, together with eye size, seems to reflect foraging differences, with species that forage from high altitudes (large eagles and large Old World vultures) having higher spatial resolution [7,8] compared with species that forage at low altitudes (Turkey vultures and black vultures [52]; Black kites [68]; American kestrel [67]), or species that forage from a perch (Harris's hawks [9,68]) or directly on the ground (Chimango caracara [69]). Having said this, based on the currently available data, scavengers, predators, and opportunists do not differ in their spatial resolution, despite the fact that predators have larger eyes and thicker retinas at the edge of the fovea. In mammals, it has also been found that visual acuity does not differ according to diet [70]. In a foraging context, predators, scavengers, or opportunists thus have similar capabilities to initially spot visual targets such as prey, carcasses, or conspecifics. Differences should (and do) occur after initial detection, where predators need to chase and follow their prey, a task that should be facilitated by the ability to fixate, high temporal resolution, and an enlarged binocular visual field.

3.2. Enlarged Field of View in Scavengers, Enlarged Binocularity in Predators

The space around an animal from which visual information can be extracted is defined by the visual field [3]. The visual field can be described by four main parameters: (1) the monocular field, the visual field of a single eye; (2) the binocular field, the area where both monocular fields overlap; (3) the cyclopean field, the total visual field produced by the combination of both monocular fields; and (4) the blind area, the space around the head from which visual information cannot be extracted [23]. In birds, visual fields vary substantially across species with different ecology [3]. For instance, while the Puna ibis *Plegadis ridwayi* (tactile forager) and the Northern bald ibis *Geronticus eremita* (visual forager) are closely related, their visual fields differ significantly, with a broader frontal binocular field in the Northern bald ibis [71]. This illustrates the trade-off between the requirement for a broader frontal binocular field for visual control of the beak/feet and the ability to gain comprehensive visual coverage for predator and conspecific detection, which has also been found in closely related ducks [72] and shorebirds [73].

This trade-off is also important in raptors. For example, species that chase highly manoeuvrable prey would be expected to possess an enlarged binocular field for visual control and accurate feet position, while scavengers would need an enlarged visual coverage for the detection of conspecifics and predators (vultures can be predated by aerial predators [74,75]). Overall, the visual fields of 18 species of diurnal raptor have been studied, including nine predators, seven scavengers, and two opportunists [4]. A recent comparative study has shown that the blind spot over the head is thinner in scavengers than in predators that hunt terrestrial prey [30] (Figure 2C). The larger blind spot above the head of predators allows better prey detection by avoiding sun dazzling. By contrast, the thinner blind spot over the head, and thus the enlarged visual coverage, of scavengers allows better conspecific detection and social foraging [76]. Predators and scavengers also differ in their binocular field shape, with a more protruding binocular field in predators [30] (Figure 2C). This is probably a result of a wider and shorter bill in species that forage on mammals [77] and enlarged optic adnexa (eyelashes and ridge above the eyes), commonly found in large-eyed species [78] to avoid sun dazzling [79].

However, some Old-World vultures (especially large species) also possess enlarged binocular fields and blind spots above the head, such as *Gyps* vultures [79] and the White-headed vulture [80], which is one of the only vulture species that have been observed to possess hunting behaviour [13]. The large blind spot above the head in *Gyps* vultures has been suggested to increase the risk of collision with wind turbines, because those vultures have a blind sector in the direction of their travel when foraging on the wing [81].

4. A Lack of Knowledge of Visual Abilities of Raptors, Especially of Scavengers

Little is known about the visual abilities of raptors. For example, we currently have estimates of visual acuity for less than 2% of all described raptor species (11 of the 557 raptors species [82]). This is even more true for scavengers. While visual acuity, and the organisation of the visual fields and/or the retina, has been assessed in an approximately equal number of predatory and scavenging raptor species, other visual aptitudes have been only studied in predatory species. In the following sections, I highlight the important gaps of the scavengers' visual abilities.

4.1. Contrast Sensitivity

Contrast sensitivity, which is a measure of how much a pattern must vary in contrast to be seen, has been estimated in only three predatory diurnal raptors: the wedge-tailed eagle (13.6 [83]), the American kestrel (30 [67]), and the Harris's hawk (12.7 [9]). While all bird species studied so far have low contrast sensitivity (from 6 to 30 [9,84]) compared with mammals (e.g., 100 in the cat [85]), it would be interesting to see whether scavengers differ from predators. In the American kestrel, the contrast sensitivity of a non-stationary (reversed) pattern (vs. stationary pattern) is higher [67]. In humans, contrast sensitivity increases significantly with a higher speed of movement [86]. Both results suggest a better detection of moving versus stationary objects. Predatory raptors may be better at detect moving targets, such as active prey, as opposed to stationary targets, such as carcasses. In contrast, because scavengers forage almost exclusively on carrion, the ability to rapidly detect stationary objects (to overcome potential competition with predators that scavenge, or other scavenging species) may be more important in these raptors. However, other non-raptorial species that seek stationary items (e.g., Common starling *Sturnus vulgaris*, Japanese quail *Coturnix japonica*, and Rock dove *Columba livia*) have similar contrast sensitivity to predators [84]. Furthermore, it has been shown that the first raptors to arrive at a carcass are often not vultures, but eagles [87], which would seem to counter the hypothesis that vultures should have higher contrast sensitivity for stationary objects. Social information is also essential for scavengers [88] and vultures also use moving objects (conspecifics) to find their food. Social information significantly facilitates foraging success in vultures [89], potentially because detecting conspecifics (moving objects) is also easier for scavengers than detecting carcasses (stationary objects). In order to better understand if different foraging tactics reflect different contrast sensitivity abilities, and if detection abilities by scavengers are improved by non-stationary objects, there is an urgent need to study scavenger species in detail.

4.2. Temporal Resolution of Vision

While spatial resolution has been studied in some raptors, including both predatory and scavenging species, the temporal resolution of vision (assessed as flicker fusion frequency, the frequency at which an intermittent light stimulus appears to be steady) has only been estimated in three predatory species, the Peregrine falcon *Falco peregrinus* (up to 125 Hz), the Saker falcon *Falco cherrug* (up to 102 Hz), and the Harris's hawk (up to 78 Hz) [90]. It has been argued that high temporal resolution should confer a selective advantage for fast-flying and manoeuvring species seeking for fast-moving prey [91]. While other species seeking stationary items, such as seeds, have high temporal resolution (e.g., up to 105 Hz in the Domestic chicken *Gallus domesticus* [92]), this seems to be confirmed in raptors, with the Peregrine falcon, the fastest animal in the world when diving on fast-moving prey [93], having the highest temporal resolution of vision among the three tested species [90]. Scavengers, especially vultures, are known to forage for carrion mainly using soaring flights [12], where rapid vision should not be necessary. However, high temporal resolution would benefit species with high spatial resolution (like large vultures) by reducing motion-induced blur [94]. Therefore, studying the temporal resolution of a broader range of raptor species and in particular some scavengers will be important to understand whether relatively high temporal resolution is common among diurnal raptors or restricted to species with predatory habits.

4.3. Colour Vision

Birds commonly use colours for discriminating object of interest, such as food, mates, or predators [95]. Most birds have a tetrachromatic vision system [96]. An important variation is found in the SWS1 cone pigment, which in birds can exist as either (1) a UV (ultraviolet) pigment or (2) a violet (V) pigment [97].

There is growing evidence that most raptors are unable to detect UV lights. Lind et al. (2013) measured the transmittance of the ocular media (cornea, lens, vitreous), which sets the limit of UV sensitivity, in four predatory raptors (the common buzzard *Buteo buteo*, the European sparrowhawks *Accipiter nisus*, the red kite *Milvus milvus*, and the Common kestrel *Falco tinnunculus*), and found UV cues are unlikely to provide a reliable visual signal to hunting raptors [98]. This is in contradiction with older behavioural studies, suggesting that vole scent marks are detectable in UV light by the Common kestrel [99,100]. Ödeen and Håstad (2013) published a study on the molecular biology of opsin genes in three other raptor species (the Turkey vulture, Pacific baza *Aviceda subcristata*, and Mississippi kite *Ictinia mississippiensis*) and found that these species have violet (but not UV) sensitive SWS1 cones [97]. Consequently, they suggested that Falconiformes and Accipitriformes cannot see UV. In 2016, Wu et al. (2016) studied the presence/absence of cone opsins in 11 additional diurnal raptor species (4 Falconiformes and 7 Accipitriformes) and found that SWS1 cone opsins were present in all species (without differentiating between violet-sensitive or UV-sensitive subtypes), except in two Accipitriformes: the Cinereous vulture *Aegypius monachus* and the Black winged kite *Elanus caeruleus* [101]. While further studies are needed to fully understand the role of UV light in raptor foraging, there is increasing evidence that most raptors cannot see UV, except maybe the Western marsh harrier *Circus aeruginosus* (Olsson, Mitkus, and Kelber, unpublished data).

Among the aforementioned species, only two scavengers have been considered (the Turkey vulture and the Cinereous vulture). Interestingly, the spectral sensitivity of these species, which are not considered to be members of the same taxonomic order, appears to differ, with the Turkey vulture (Cathartiformes) having tetrachromatic vision, but the Cinereous vulture (Accipitriformes) only having trichromatic vision [97,101]. Furthermore, the detailed spectral sensitivity of scavengers has never been studied contrary to predators (Common kestrel, Sparrowhawk *Accipiter nisus*, and Common buzzard *Buteo buteo* [98]).

Finally, the use of colour information appears to be important at a distance in raptors, with Harris's hawk having twice as good a chromatic spatial resolution compared with humans, while achromatic vision is similar [9]. The spatial resolution of a chromatic vision has only been estimated in two bird species, Harris's hawk [9] and the non-raptorial budgerigar *Melopsittacus undulatus* [102]. While spatial resolution of the chromatic vision is systematically lower than for the achromatic channel, understanding whether scavengers can discriminate colours at a distance would be essential to understand which cues are important to these birds for detecting carrion.

5. Conclusions

While raptors are often cited as having extraordinary sight, what I have highlighted here is that we actually have scarce knowledge on the visual sense in the vast majority of diurnal raptors, and especially in scavengers. Here, I reviewed what is currently known about the functional differences in the visual abilities of predators and scavengers. Predators have a visual system adapted to predation, characterised by the following: (1) large eyes with two foveas, one located centrally for the detection of prey at distance, and one located temporally to fixate the prey at the moment of capture and (2) a visual field with an enlarged binocular field to facilitate the guidance and positioning of the feet during prey capture, and a larger blind spot above the head to avoid sun-dazzling for species chasing preys on the ground. In contrast, scavengers have a visual system adapted to carrion eating and social foraging: (1) smaller eyes (compared with body mass) with only one central fovea to spot carrion at distance and (2) an enlarged field of view to facilitate the detection of carcasses and for the accurate use of social information. Despite these differences, it is important to remember that the visual systems of predators and scavengers are also very similar in a number of ways. For example, they both have the following: (1) a classical vertebrate optical system; (2) an identical design of the retinal layers; and (3) spatial resolution that appears to be adapted to flight altitude during foraging rather than diet per se, with some species having the highest visual acuity in the Animal Kingdom.

With raptors globally suffering from population declines [82], increasing our knowledge about their sensory ecology and behaviour could offer significant benefits to conservation programs [103]. In particular, scavengers, and especially Old-World vultures, are more threatened than any of the other groups [82], with over 80% of species declining. It has been shown that understanding their visual systems may help protect them, such as understanding the high collision rate with human devices [81]. Therefore, in light of our minimal knowledge on the sensory biology of scavenging raptors, there is urgency in studying their visual capacities.

Supplementary Materials: The following are available online at http://www.mdpi.com/1424-2818/12/10/400/s1, Table S1: Information about foraging tactics, corneal diameter, axial length and body mass of raptors.

Funding: I gratefully acknowledge support from the K. & A. Wallenberg Foundation, Stockholm (Ultimate Vision) and the Swedish Research Council (2016-03298_3).

Acknowledgments: I sincerely thank A. Kelber, M. Mitkus, and T. Lisney for their proof-reading and constructive comments on this review.

Conflicts of Interest: The author declares no conflict of interest.

References

1. Stevens, M. *Sensory Ecology, Behaviour, and Evolution*; Oxford University Press: Oxford, UK, 2013.
2. Land, M.F.; Nilsson, D.-E. *Animal Eyes*; Oxford University Press: Oxford, UK, 2012.
3. Martin, G.R. *The Sensory Ecology of Birds*; Oxford University Press: Oxford, UK, 2017.
4. Mitkus, M.; Potier, S.; Martin, G.R.; Duriez, O.; Kelber, A. Raptor vision. In *Oxford Research Encyclopedia of Neuroscience*; Oxford University Press: Oxford, UK, 2018.
5. Potier, S.; Mitkus, M.; Kelber, A. Visual adaptations of diurnal and nocturnal raptors. *Semin. Cell Dev. Biol.* **2020**, in press. [CrossRef] [PubMed]
6. Potier, S. Olfaction in raptors. *Zool. J. Linn. Soc.* **2019**, *189*, 713–721. [CrossRef]

7.   Reymond, L. Spatial visual acuity of the eagle *Aquila audax*: A behavioural, optical and anatomical investigation. *Vis. Res.* **1985**, *25*, 1477–1491. [CrossRef]

8.   Fischer, A.B. Laboruntersuchungen und freilandbeobachtungen zum sehvermögen und verhalten von altweltgeiern. *Zool. Jahrb. Syst. Ökol. Geogr. Tiere* **1968**, *96*, 81–132.

9.   Potier, S.; Mitkus, M.; Kelber, A. High resolution of colour vision, but low contrast sensitivity in a diurnal raptor. *Proc. R. Soc. B Biol. Sci.* **2018**, *285*, 20181036. [CrossRef]

10. McClure, C.J.W.; Schulwitz, S.E.; Anderson, D.L.; Robinson, B.W.; Mojica, E.L.; Therrien, J.-F.; Oleyar, M.D.; Johnson, J. Commentary: Defining Raptors and Birds of Prey. *J. Raptor Res.* **2019**, *53*, 419–430. [CrossRef]

11. Jetz, W.; Thomas, G.H.; Joy, J.B.; Hartmann, K.; Mooers, A.O. The global diversity of birds in space and time. *Nature* **2012**, *491*, 444–448. [CrossRef]

12. Ferguson-Lees, J.; Christie, D.A. *Raptors of the World*; A&C Black: London, UK, 2001.

13. Murn, C. Observations of predatory behavior by white-headed vultures. *J. Raptor Res.* **2014**, *48*, 297–299. [CrossRef]

14. Kapfer, J.M.; Gammon, D.E.; Groves, J.D. Carrion-feeding by Barred Owls (*Strix varia*). *Wilson J. Ornithol.* **2011**, *123*, 646–649. [CrossRef]

15. Milchev, B.; Spassov, N. First evidence for carrion—Feeding of Eurasian Eagle-owl (*Bubo bubo*) in Bulgaria. *Ornis Hung.* **2017**, *25*, 58–69. [CrossRef]

16. Mori, E.; Menchetti, M.; Dartora, F. Evidence of carrion consumption behaviour in the Long-eared Owl *Asio otus* (Linnaeus, 1758) (Aves: Strigiformes: Strigidae). *Ital. J. Zool.* **2014**, *81*, 471–475. [CrossRef]

17. Allen, M.L.; Taylor, A.P. First record of scavenging by a Western Screech-Owl (*Megascops kennicottii*). *Wilson J. Ornithol.* **2013**, *125*, 417–419. [CrossRef]

18. Wilman, H.; Belmaker, J.; Simpson, J.; de la Rosa, C.; Rivadeneira, M.M.; Jetz, W. EltonTraits 1.0: Species-level foraging attributes of the world's birds and mammals: Ecological Archives E095–179. *Ecology* **2014**, *95*, 2027. [CrossRef]

19. Walls, G.L. *The Vertebrate Eye and Its Adaptive Radiation*; Cranbrook Institute of Science : Bloomfield Hills, MI, USA; Cranbrook Press: Toowoomba City, QLD, Australia, 1942.

20. Brooke, M.L.; Hanley, S.; Laughlin, S.B. The scaling of eye size with body mass in birds. *Proc. R. Soc. B Biol. Sci.* **1999**, *266*, 405–412. [CrossRef]

21. Kiltie, R.A. Scaling of visual acuity with body size in mammals and birds. *Funct. Ecol.* **2000**, *14*, 226–234. [CrossRef]

22. Niven, J.E.; Laughlin, S.B. Energy limitation as a selective pressure on the evolution of sensory systems. *J. Exp. Biol.* **2008**, *211*, 1792–1804. [CrossRef]

23. Martin, G.R. Visual fields and their functions in birds. *J. Ornithol.* **2007**, *148*, 547–562. [CrossRef]

24. Potier, S.; Mitkus, M.; Bonadonna, F.; Duriez, O.; Isard, P.-F.; Dulaurent, T.; Mentek, M.; Kelber, A. Eye size, fovea, and foraging ecology in accipitriform raptors. *Brain Behav. Evol.* **2017**, *90*, 232–242. [CrossRef]

25. Zelenitsky, D.K.; Therrien, F.; Ridgely, R.C.; McGee, A.R.; Witmer, L.M. Evolution of olfaction in non-avian theropod dinosaurs and birds. *Proc. R. Soc. B Biol. Sci.* **2011**, *279*, 3625–3634. [CrossRef] [PubMed]

26. Bang, B.G.; Cobb, S. The size of the olfactory bulb in 108 species of birds. *Auk* **1968**, *85*, 55–61.

27. Corfield, J.R.; Price, K.; Iwaniuk, A.N.; Gutiérrez-Ibáñez, C.; Birkhead, T.; Wylie, D.R. Diversity in olfactory bulb size in birds reflects allometry, ecology, and phylogeny. *Front. Neuroanat.* **2015**, *9*, 102. [CrossRef] [PubMed]

28. Tucker, V.A. The deep fovea, sideways vision and spiral flight paths in raptors. *J. Exp. Biol.* **2000**, *203*, 3745–3754. [PubMed]

29. Potier, S.; Bonadonna, F.; Martin, G.R.; Isard, P.-F.; Dulaurent, T.; Mentek, M.; Duriez, O. Visual configuration of two species of Falconidae with different foraging ecologies. *Ibis* **2017**, *160*, 54–61. [CrossRef]

30. Potier, S.; Duriez, O.; Cunningham, G.B.; Bonhomme, V.; O'Rourke, C.; Fernández-Juricic, E.; Bonadonna, F. Visual field shape and foraging ecology in diurnal raptors. *J. Exp. Biol.* **2018**, *221*, jeb177295. [CrossRef]

31. Wickham, H. *ggplot2: Elegant Graphics for Data Analysis. Version 3.3.0*; Springer-Verlag: New York, NY, USA, 2016.

32. Ho, L.S.T.; Ane, C. phylolm: A linera-time algorithm for Gaussian and non-Gaussian trait evolution models. *Syst. Biol.* **2014**, *63*, 63–408.

33. Revell, L.J. phytools: An R package for phylogenetic comparative biology (and other things). *Methods Ecol. Evol.* **2012**, *3*, 217–223. [CrossRef]

34. Orme, D.; Freckleton, R.; Thomas, G.; Petzoldt, T.; Fritz, S.; Isaac, N.; Pearse, W. The caper: Comparative Analysis of Phylogenetics and Evolution in R. Available online: https://cran.r-project.org/web/packages/caper/vignettes/ (accessed on 24 September 2020).

35. Zeileis, A.; Hothorn, T. lmtest: Diagnostic Checking in Regression Relationships. Available online: http://pkg.cs.ovgu.de/LNF/i386/5.10/R/LNFr-lmtest/reloc/R-2.10/library/lmtest/ (accessed on 24 September 2020).

36. Kassambara, A. ggpubr: "ggplot2" based publication ready plots. R Package Version 0.2.5. 2020. Available online: https://CRAN.R-project.org/package=ggpubr (accessed on 24 September 2020).

37. Wickham, H. plyr: The split-apply-combine strategy for data analysis. *J. Stat. Soft.* **2011**, *40*, 1–29. [CrossRef]

38. Dunning, J.B., Jr. *CRC Handbook of Avian Body Masses*; CRC Press: Boca Raton, FL, USA, 2007.

39. Hwang, J.; Kang, S.; Seok, S.; Ahmed, S.; Yeon, S. Ophtalmic findings in cinereous vultures (*Aegypius monachus*). *Vet. Ophthalmol.* **2020**, *23*, 314–324. [CrossRef]

40. Ritland, S.M. *The Allometry of the Vertebrate Eye*; Department of Biology, University of Chicago: Chicago, IL, USA, 1984.

41. Spiegel, O.; Getz, W.M.; Nathan, R. Factors influencing foraging search efficiency: Why do scarce lappet-faced vultures outperform ubiquitous white-backed vultures? *Am. Nat.* **2013**, *181*, 102–115. [CrossRef] [PubMed]

42. Martin, G.R. Schematic eye models in vertebrates. In *Progress in Sensory Physiology*; Springer: Berlin/Heidelberg, Germany, 1984; pp. 43–81.

43. Miller, W.H. Ocular optical filtering. In *Comparative Physiology and Evolution of Vision in Invertebrates*; Springer: Berlin/Heidelberg, Germany, 1979; pp. 69–143.

44. Glasser, A.; Howland, H.C. A history of studies of visual accommodation in birds. *Q. Rev. Biol.* **1996**, *71*, 475–509. [CrossRef] [PubMed]

45. Ott, M. Visual accommodation in vertebrates: Mechanisms, physiological response and stimuli. *J. Comp. Physiol. A* **2006**, *192*, 97–111. [CrossRef] [PubMed]

46. Glasser, A.; Pardue, M.T.; Andison, M.E.; Sivak, J.G. A behavioral study of refraction, corneal curvature, and accommodation in raptor eyes. *Can. J. Zool.* **1997**, *75*, 2010–2020. [CrossRef]

47. Sivak, J.G.; Hildebrand, T.; Lebert, C. Magnitude and rate of accommodation in diving and nondiving birds. *Vis. Res.* **1986**, *25*, 925–933. [CrossRef]

48. Schaeffel, F.; Howland, H.C.; Farkas, L. Natural accommodation in the growing chicken. *Vis. Res.* **1986**, *26*, 1977–1993. [CrossRef]

49. Rochon-Duvigneaud, A. *Les yeux et la vision des vertébrés*; Masson Paris: Paris, France, 1943.

50. Fite, K.V.; Rosenfield-Wessels, S. A comparative study of deep avian foveas. *Brain Behav. Evol.* **1975**, *12*, 97–115. [CrossRef]

51. Inzunza, O.; Bravo, H.; Smith, R.L.; Angel, M. Topography and morphology of retinal ganglion cells in Falconiforms: A study on predatory and carrion-eating birds. *Anat. Rec.* **1991**, *229*, 271–277. [CrossRef]

52. Lisney, T.J.; Stecyk, K.; Kolominsky, J.; Graves, G.R.; Wylie, D.R.; Iwaniuk, A.N. Comparison of eye morphology and retinal topography in two species of new world vultures (Aves: Cathartidae). *Anat. Rec.* **2013**, *296*, 1954–1970. [CrossRef]

53. Bringmann, A. Structure and function of the bird fovea. *Anat. Histol. Embryol.* **2019**, *48*, 177–200. [CrossRef]

54. Oehme, H. Vergleichende untersuchungen an greifvogelaugen. *Z. Für Morphol. Und Ökologie Der Tiere* **1964**, *53*, 618–635. [CrossRef]

55. Coimbra, J.P.; Collin, S.P.; Hart, N.S. Variations in retinal photoreceptor topography and the organization of the rod-free zone reflect behavioral diversity in Australian passerines. *J. Comp. Neurol.* **2015**, *523*, 1073–1094. [CrossRef]

56. Bringmann, A.; Syrbe, S.; Görner, K.; Kacza, J.; Francke, M.; Wiedemann, P.; Reichenbach., A. The primate fovea: Structure, function and development. *Prog. Retin. Eye Res.* **2018**, *66*, 49–84. [CrossRef] [PubMed]

57. Walls, G.L. Significance of the foveal depression. *Arch. Ophthalmol.* **1937**, *18*, 912–919. [CrossRef]

58. Snyder, A.W.; Miller, W.H. Telephoto lens system of falconiform eyes. *Nature* **1979**, *275*, 127–129. [CrossRef] [PubMed]

59. Potier, S.; Mitkus, M.; Lisney, T.J.; Isard, P.-F.; Dulaurent, T.; Mentek, M.; Cornette, R.; Schikorski, D.; Kelber, A. Inter-individual differences in foveal shape in a scavenging raptor, the black kite *Milvus migrans. Sci. Rep.* **2020**, *10*, 1–12. [CrossRef]

60. Reymond, L. Spatial visual acuity of the falcon, *Falco berigora*: A behavioural, optical and anatomical investigation. *Vis. Res.* **1987**, *27*, 1869–1874. [CrossRef]

61. Pumphrey, P.J. The theory of the fovea. *J. Exp. Biol.* **1948**, *25*, 299–312.

62. Ruggeri, M.; Major, J.C.; McKeown, C.; Knighton, R.W.; Puliafito, C.A.; Jiao, S. Retinal structure of birds of prey revealed by ultra-high resolution spectral-domain optical coherence tomography. *Investig. Ophthalmol. Vis. Sci.* **2010**, *51*, 5789–5795. [CrossRef]

63. Martin, G.R. What is binocular vision for? A birds' eye view. *J. Vis.* **2009**, *9*, 14. [CrossRef]

64. Moroney, M.K.; Pettigrew, J.D. Some observations on the visual optics of kingfishers (Aves, Coraciformes, Alcedinidae). *J. Comp. Physiol. A* **1987**, *160*, 137–149. [CrossRef]

65. Tyrrell, L.P.; Fernández-Juricic, E. The hawk-eyed songbird: Retinal morphology, eye shape, and visual fields of an aerial insectivore. *Am. Nat.* **2017**, *189*, 709–717. [CrossRef]

66. Fox, R.; Lehmkuhle, S.W.; Westendorf, D.H. Falcon visual acuity. *Science* **1976**, *192*, 263–265. [CrossRef] [PubMed]

67. Hirsch, J. Falcon visual sensitivity to grating contrast. *Nature* **1982**, *300*, 57–58. [CrossRef]

68. Potier, S.; Bonadonna, F.; Kelber, A.; Martin, G.R.; Isard, P.-F.; Dulaurent, T.; Duriez, O. Visual abilities in two raptors with different ecology. *J. Exp. Biol.* **2016**, *219*, 2639–2649. [CrossRef]

69. Potier, S.; Bonadonna, F.; Kelber, A.; Duriez, O. Visual acuity in an opportunistic raptor, the chimango caracara (*Milvago chimango*). *Physiol. Behav.* **2016**, *157*, 125–128. [CrossRef] [PubMed]

70. Veilleux, C.C.; Kirk, E.C. Visual acuity in mammals: Effects of eye size and ecology. *Brain Behav. Evol.* **2014**, *84*, 43–53. [CrossRef]

71. Martin, G.R.; Portugal, S.J. Differences in foraging ecology determine variation in visual fields in ibises and spoonbills (Threskiornithidae). *Ibis* **2011**, *153*, 662–671. [CrossRef]

72. Martin, G.R.M.; Jarrett, N.; Williams, M. Visual fields in Blue Ducks *Hymenolaimus malacorhynchos* and Pink-eared Ducks *Malacorhynchus membranaceus*: Visual and tactile foraging. *Ibis* **2007**, *149*, 112–120. [CrossRef]

73. Martin, G.R.; Piersma, T. Vision and touch in relation to foraging and predator detection: Insightful contrasts between a plover and a sandpiper. *Proc. R. Soc. B Biol. Sci.* **2009**, *276*, 437–445. [CrossRef]

74. Boal, C. Golden Eagle predation of an adult Turkey Vulture. *Bull. Tex. Ornithol. Soc.* **2015**, *48*, 53–55.

75. Thompson, L.J.; Clemence, L.; Clemence, B.; Goosen, D. Nestling White-backed Vulture (*Gyps africanus*) eaten by a Verreaux's Eagle (*Aquila verreauxii*) at a nest occupied for a record 21 years. *Vulture News* **2018**, *74*, 24–30. [CrossRef]

76. Fernández-Juricic, E.; Erichsen, J.T.; Kacelnik, A. Visual perception and social foraging in birds. *Trends Ecol. Evol.* **2004**, *19*, 25–31. [CrossRef] [PubMed]

77. Slagsvold, T.; Sonerud, G.A.; Grønlien, H.E.; Stige, L.C. Prey handling in raptors in relation to their morphology and feeding niches. *J. Avian Biol.* **2010**, *41*, 488–497. [CrossRef]

78. Martin, G.R.; Coetzee, H.C. Visual fields in hornbills: Precision-grasping and sunshades. *Ibis* **2004**, *146*, 18–26. [CrossRef]

79. Martin, G.R.; Katzir, G. Sun shades and eye size in birds. *Brain Behav. Evol.* **2000**, *56*, 340–344. [CrossRef] [PubMed]

80. Portugal, S.J.; Murn, C.P.; Martin, G.R. White-headed Vulture *Trigonoceps occipitalis* shows visual field characteristics of hunting raptors. *Ibis* **2017**, *159*, 463–466. [CrossRef]

81. Martin, G.R.; Portugal, S.J.; Murn, C.P. Visual fields, foraging and collision vulnerability in Gyps vultures. *Ibis* **2012**, *154*, 626–631. [CrossRef]

82. McClure, C.J.W.; Westrip, J.R.S.; Johnson, J.A.; Schulwitz, S.E.; Virani, M.Z.; Davies, R.; Symes, A.; Wheatley, H.; Thorstrom, R.; Amar, A. State of the world's raptors: Distributions, threats, and conservation recommendations. *Biol. Conserv.* **2018**, *227*, 390–402. [CrossRef]

83. Reymond, L.; Wolfe, J. Behavioural determination of the contrast sensitivity function of the eagle *Aquila audax*. *Vis. Res.* **1981**, *21*, 263–271. [CrossRef]

84. Ghim, M.M.; Hodos, W. Spatial contrast sensitivity of birds. *J. Comp. Physiol. A* **2006**, *192*, 523–534. [CrossRef]

85. Campbell, F.W.; Maffei, M.; Piccolino, M. The contrast sensitivity of the cat. *J. Physiol.* **1973**, *229*, 719–731. [CrossRef]

86. Virsu, V.; Rovamo, J.; Laurinen, P.; Näsänen, R. Temporal contrast sensitivity and cortical magnification. *Vis. Res.* **1982**, *22*, 1211–1217. [CrossRef]

87. Kane, A.; Jackson, A.L.; Ogada, D.L.; Monadjem, A.; McNally, L. Vultures acquire information on carcass location from scavenging eagles. *Proc. R. Soc. B Biol. Sci.* **2014**, *281*, 20141072. [CrossRef]

88. Cortés-Avizanda, A.; Jovani, R.; Donázar, J.A.; Grimm, V. Bird sky networks: How do avian scavengers use social information to find carrion? *Ecology* **2014**, *95*, 1799–1808. [CrossRef] [PubMed]

89. Jackson, A.L.; Ruxton, G.D.; Houston, D.C. The effect of social facilitation on foraging success in vultures: A modelling study. *Biol. Lett.* **2008**, *4*, 311–313. [CrossRef] [PubMed]

90. Potier, S.; Lieuvin, M.; Pfaff, M.; Kelber, A. How fast can raptors see? *J. Exp. Biol.* **2020**, *223*, jeb209031. [CrossRef] [PubMed]

91. Boström, J.E.; Dimitrova, M.; Canton, C.; Håstad, O.; Qvarnström, A.; Ödeen, A. Ultra-rapid vision in birds. *PLoS ONE* **2016**, *11*, e0151099. [CrossRef]

92. Nuboer, J.F.W.; Coemans, M.; Vos, J.J. Artificial lighting in poultry houses: Do hens perceive the modulation of fluorescent lamps as flicker? *Br. Poult. Sci.* **1992**, *33*, 123–133. [CrossRef]

93. Tucker, V.A. Gliding flight: Speed and acceleration of ideal falcons during diving and pull out. *J. Exp. Biol.* **1998**, *201*, 403–414.

94. Srinivasan, M.V.; Bernard, G.D. The effect of motion on visual acuity of the compound eye: A theoretical analysis. *Vis. Res.* **1975**, *15*, 515–525. [CrossRef]

95. Kelber, A. Bird colour vision—From cones to perception. *Curr. Opin. Behav. Sci.* **2019**, *30*, 34–40. [CrossRef]

96. Hart, N.S.; Hunt, D.M. Avian visual pigments: Characteristics, spectral tuning, and evolution. *Am. Nat.* **2007**, *169*, S7–S26. [CrossRef]

97. Ödeen, A.; Håstad, O. The phylogenetic distribution of ultraviolet sensitivity in birds. *BMC Evol. Biol.* **2013**, *13*, 36. [CrossRef] [PubMed]

98. Lind, O.; Mitkus, M.; Olsson, P.; Kelber, A. Ultraviolet sensitivity and colour vision in raptor foraging. *J. Exp. Biol.* **2013**, *216*, 1819–1826. [CrossRef]

99. Viitala, J.; Korpimäki, E.; Palokangas, P.; Koivula, M. Attraction of kestrels to vole scent marks visible in ultraviolet light. *Nature* **1995**, *373*, 425–427. [CrossRef]

100. Koivula, M.; Viitala, J.; Koipimaki, E. Kestrels prefer scent marks according to species and reproductive status of voles. *Ecoscience* **1999**, *6*, 415–420. [CrossRef]

101. Wu, Y.; Hadly, E.A.; Teng, W.; Hao, Y.; Liang, W.; Liu, Y.; Wang, H. Retinal transcriptome sequencing sheds light on the adaptation to nocturnal and diurnal lifestyles in raptors. *Sci. Rep.* **2016**, *6*, 1–12. [CrossRef] [PubMed]

102. Lind, O.; Kelber, A. The spatial tuning of achromatic and chromatic vision in budgerigars. *J. Vis.* **2011**, *11*, 1–19. [CrossRef] [PubMed]

103. Dominoni, D.M.; Halfwerk, W.; Baird, E.; Buxton, R.T.; Fernández-Juricic, E.; Fristrup, K.M.; McKenna, M.F.; Mennitt, D.J.; Perkin, E.K.; Seymoure, B.M.; et al. Why conservation biology can benefit from sensory ecology. *Nat. Ecol. Evol.* **2020**, *4*, 502–511. [CrossRef]

Publisher's Note: MDPI stays neutral with regard to jurisdictional claims in published maps and institutional affiliations.

Review

The Role of Carrion in the Landscapes of Fear and Disgust: A Review and Prospects

Marcos Moleón [1,*] and José A. Sánchez-Zapata [2]

[1] Department of Zoology, University of Granada, 18071 Granada, Spain
[2] Department of Applied Biology, University Miguel Hernández, 03202 Elche, Spain; toni@umh.es
* Correspondence: mmoleonpaiz@hotmail.com or mmoleon@ugr.es

Abstract: Animal behavior is greatly shaped by the 'landscape of fear', induced by predation risk, and the equivalent 'landscape of disgust', induced by parasitism or infection risk. However, the role that carrion may play in these landscapes of peril has been largely overlooked. Here, we aim to emphasize that animal carcasses likely represent ubiquitous hotspots for both predation and infection risk, thus being an outstanding paradigm of how predation and parasitism pressures can concur in space and time. By conducting a literature review, we highlight the manifold inter- and intra-specific interactions linked to carrion via predation and parasitism risks, which may affect not only scavengers, but also non-scavengers. However, we identified major knowledge gaps, as reviewed articles were highly biased towards fear, terrestrial environments, vertebrates, and behavioral responses. Based on the reviewed literature, we provide a conceptual framework on the main fear- and disgust-based interaction pathways associated with carrion resources. This framework may be used to formulate predictions about how the landscape of fear and disgust around carcasses might influence animals' individual behavior and ecological processes, from population to ecosystem functioning. We encourage ecologists, evolutionary biologists, epidemiologists, forensic scientists, and conservation biologists to explore the promising research avenues associated with the scary and disgusting facets of carrion. Acknowledging the multiple trophic and non-trophic interactions among dead and live animals, including both herbivores and carnivores, will notably improve our understanding of the overlapping pressures that shape the landscape of fear and disgust.

Keywords: carcass; confrontational scavenging; disease risk; facultative scavenger; landscape of peril; marine ecosystems; parasite risk; predator risk; terrestrial ecosystems

Citation: Moleón, M.; Sánchez-Zapata, J.A. The Role of Carrion in the Landscapes of Fear and Disgust: A Review and Prospects. *Diversity* **2021**, *13*, 28. https://doi.org/10.3390/d13010028

Received: 3 December 2020
Accepted: 11 January 2021
Published: 13 January 2021

Publisher's Note: MDPI stays neutral with regard to jurisdictional claims in published maps and institutional affiliations.

1. Introduction

Recently, Buck et al. [1] and Weinstein et al. [2] formalized a correspondence between predator and parasite avoidance behaviors. They argued how infection risk must determine a three-dimensional 'landscape of disgust' equivalent to the 'landscape of fear' induced by predation risk (defined by [1] as "the relative levels of predation risk experienced by a prey individual, represented as peaks and valleys on the landscape"). In this way, animal behavior is largely shaped by perceived risk (from either predators or parasites), leading to high-risk sites avoidance and preference for low-risk patches [1,2]. Either jointly or independently, these natural enemies may lead to fitness costs in their victims through physiological (e.g., chronic stress [3]) and behavioral (e.g., changes in habitat preferences [4]) effects. Strikingly, these physiological responses and behavioral decisions not only result from direct encounters with enemies, but frequently rely on indirect cues linked to risk situations or resources, regardless of actual presence of predators or parasites (e.g., [5]). Thus, through inducing fear and disgust, predators and parasites lead to a pervasive 'landscape of peril' [6] that may indirectly affect individuals and populations, as well as communities and ecosystems via cascading effects [1,2,6].

If we delve into this integrative, general view, many scary (i.e., related to the landscape of fear) and disgusting (i.e., related to the landscape of disgust) facets of animal carcasses can be inferred. In doing so, it rapidly comes to light that carrion probably represents ubiquitous—as it is produced in all biomes—hotspots for both predation and infection risk. On one hand, predators of different trophic levels are usually attracted to carcasses worldwide [7], which may lead to predation risk to not only herbivores (e.g., [8]), but also subordinate predators (e.g., [9]). On the other hand, carrion has long been considered a prominent source of pathogens that can put scavenging animals (e.g., [10]) and other species that may be present at carcass sites (e.g., [11]) at risk. For instance, the behavioral and cognitive repertoire of modern humans are probably shaped in part by the exposure of earliest hominins to the risks of being killed or injured while scavenging large herbivore carcasses and acquiring parasites when consuming a decaying piece of meat [12–14]. Pressures associated with predation and parasitism risk at carrion resources seem to be so pervasive that even some plants have taken advantage of them. In particular, species of genera *Rafflesia*, *Aristolochia*, and *Helicodiceros*, among others, could use thanatosis (i.e., olfactory feigning of carrion) to not only attract pollinators, but also deter herbivores, especially during the flowering period [15].

However, we are just starting to uncover the manifold ecological and evolutionary ramifications of carrion within the context of predation and parasitism risks. Research on this topic is especially needed given the ongoing global environmental change. Understanding how animals thrive in the changing landscape of peril associated with carrion could provide important insights for the conservation of threatened scavengers. Moreover, studying how animals behave around carcass sites could reveal key findings of veterinary and epidemiological interest, which is particularly relevant in the current context of zoonotic diseases [16].

Our general aim is to examine the role that carrion plays in the landscapes of fear and disgust, which has been largely overlooked in the scientific literature despite its crucial eco-evolutionary, epidemiological, and management implications. Through a bibliographic review, we will identify the main ways in which carrion may be scary and disgusting, namely the principal interaction pathways between carcasses and their visitors (both carnivores and herbivores) that expose the former to predators and parasites at carcass sites. Here, predators and scavengers are defined as gatherers and miners, respectively, of live animals [17], with parasites including macroparasites, protists, fungi, bacteria, and viruses [18]. Then, we will determine the main knowledge gaps and provide ideas for future investigation on this emerging and highly promising research topic.

2. Material and Methods

Following guidelines provided by Haddaway et al. [19], we used the Web of Science to conduct a systematic review of the scientific literature on the landscapes of fear and disgust associated with carrion. Specifically, using the "Topic search" (i.e., title, abstract, and keywords), we searched for "articles" appearing prior to November 2020 that included the following combinations of terms: "landscape of fear" OR "fear" OR "predat* risk" AND "carrion" OR "carcass" OR "scaveng*"; and "landscape of disgust" OR "disgust" OR "parasit* risk" OR "parasit* avoidance" OR "disease risk" OR "disease avoidance" OR "infection risk" OR "infection avoidance" AND "carrion" OR "carcass" OR "scaveng*" (Appendix A). We further restricted our search in a two-steps process (e.g., [20]). First, title and abstract were screened to ensure we only included empirical studies dealing with the general topic of this review. Second, we read the full content of the selected articles, excluding articles mentioning only superficially in the introduction and discussion the searched terms, e.g., to motivate the study or suggest future research needs. Through this procedure, we obtained 26 articles. Then, we used Google Scholar to identify additional papers, restricting the search to the first 30 papers for each combination of terms. This complementary search provided 26 articles not identified previously in the Web of Science. In total, we obtained a final set of 52 articles for in-depth review (see References A1 for a complete list of reviewed references), which we consider sufficient to infer global patterns of research effort and relative differences among distinct interaction pathways.

We divided the selected articles in two main groups, depending on whether they were concerned with fear or disgust. Then, from each article, we obtained the following information. First, we extracted general data: year of publication; ecosystem under study (*terrestrial, coastal, marine, freshwater*); geographic location (i.e., the country where the study was conducted); period under study (*prehistorical, historical*); animals involved in the study (i.e., scavengers and non-scavengers that may feed or inspect around carcasses; *vertebrates, invertebrates*); the type of effects—studied, detected, or presumed—that the fear or disgust exerted on such species (*behavioral, physiological, demographic, non-specified*), and study design (*field study, observational; field study, experimental; field study, quasi-experimental; mesocosm experiment; other*). "Quasi-experimental" studies refer to those in which carcasses or artificial nests were placed in locations selected by the researchers, but no other condition was manipulated. For each field and mesocosm study, we also recorded the observation method (*direct observation, camera-trap, other*). Second, we identified the ways in which carrion and visitors to carrion sites could lead to predation or parasitism risk to these or other visitors. This, along with other key reviews on the topic (e.g., [21–27]), was the basis to elaborate a conceptual framework on the main interaction pathways around carrion resources that are related to fear and disgust. In the latter case, we distinguished consumptive (trophic) and non-consumptive (non-trophic) processes. We represented the framework separately for herbivore and carnivore carcasses, given that their decomposition process, persistence time in the environment, and associated risks are neatly different [28]. Third, we recorded the number of articles selected in the review that were related to each pathway, with the aim of identifying major knowledge gaps that could be key targets for future research.

3. Results and Discussion

3.1. General Results

We obtained more articles concerned with fear (75%, $n = 39$) than with disgust (25%, $n = 13$). No article empirically explored simultaneously both types of risk. Moreover, scientists became interested later on disgust than on fear, according to the year of publication of these articles (Figure 1A). Most articles on fear, and all articles on disgust, involved terrestrial ecosystems or mesocosms. Thus, representation of articles dealing with aquatic environments was scarce (Figure 1B). Most articles focused on present-day assemblages, while all studies concerned with prehistorical times were related to the predation (mainly) and infection risk faced at carcass sites by early hominins (Figure 1C). There were more articles studying vertebrates than invertebrates (Figure 1D), with the latter being mainly associated with marine and freshwater systems. Effects on visitors to carrion sites, as recorded in the reviewed articles, were mostly behavioral; only a few articles involved demographic effects (in all cases, related to bird nest predation), and none explored physiological effects on visitor species (Figure 1E). Studies on predation risk were mostly observational and quasi-experimental, though experimental approaches (all in mesocosm systems) were frequently used to assess parasitism risk. In addition, the intentional deployment of carcasses (and artificial nests) was normally associated with the use of camera-traps for monitoring bait use by animals, especially in disgust-related studies (Figure 1F). Most present-day studies were conducted in Australia, USA, and Europe, while all studies on early hominins were done with material from eastern Africa (Figure 1G).

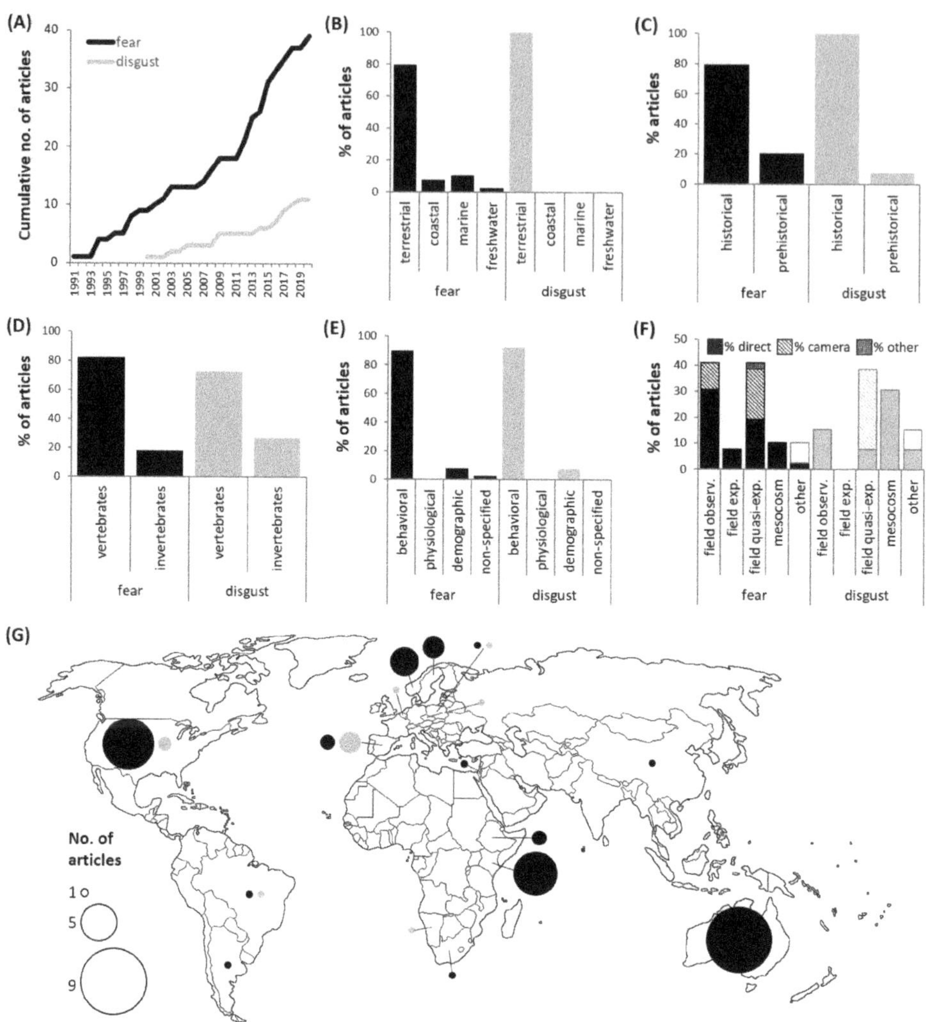

Figure 1. Distribution of reviewed articles according to publication year (**A**), ecosystem (**B**), period (**C**), type of animals at (potential) risk (**D**), effects of the risks on such animals (**E**), study design (different bars) and observation method (within bars) (**F**), and geographic location (**G**). See main text for details.

3.2. Interaction Pathways and Research Effort

Figure 2 shows our conceptual model for the main fear- and disgust-based interaction pathways associated with carcasses of carnivore and herbivore species. In general, while animals may be mainly disgusted by carcasses themselves, scare is genuinely related to other animals that may be attracted to the carcass for any reason or present by chance in its vicinities.

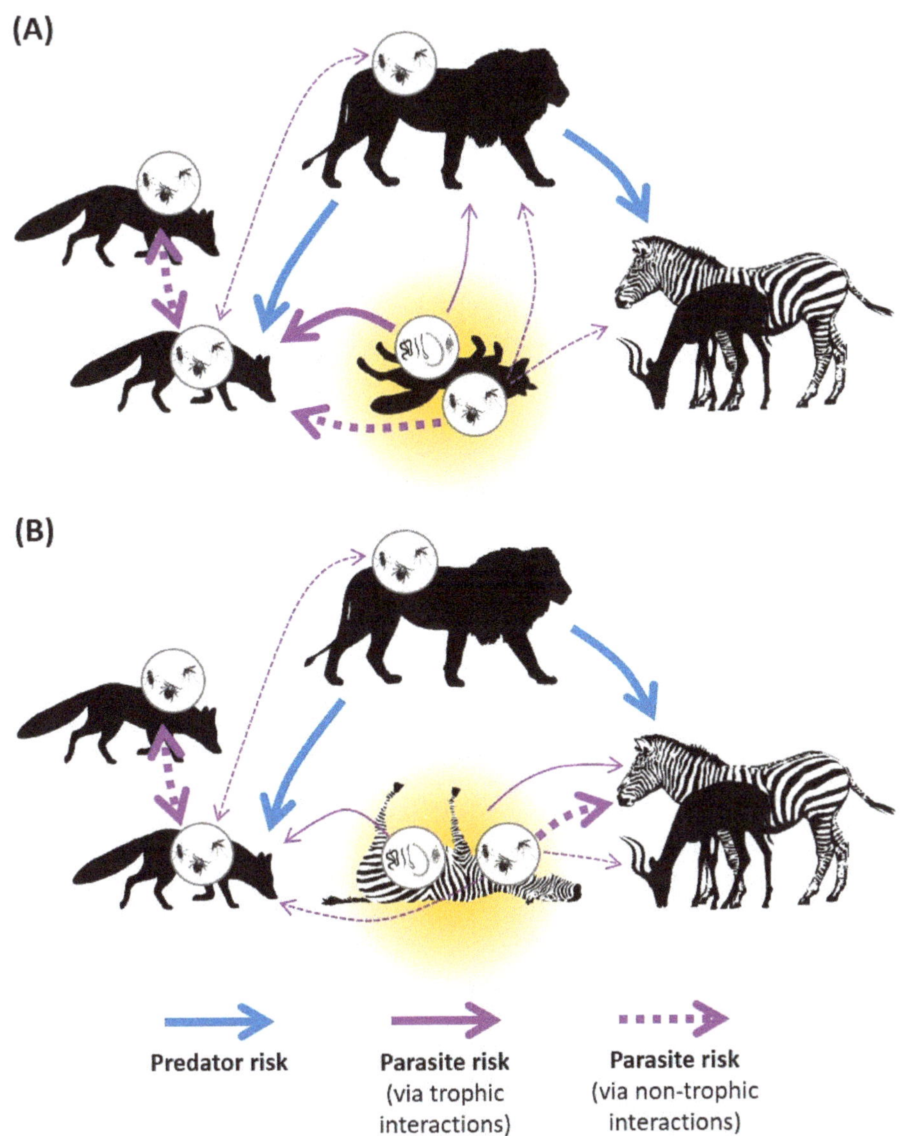

Figure 2. Animal carcasses may be both disgusting and scary. However, the parasite and predator risks vary widely according to the species identity of the carcass and its visitors. (**A**) Carnivores visiting carnivore carcasses may face both predation and infection risk. Parasites may be transmitted to carnivores by either the carcass or other carnivore visitors, with the probability of transmission being proportional to their phylogenetic relationship [29,30]. (**B**) Herbivore carcasses are relatively safe for carnivores in terms of direct parasite transmission (at least, regarding direct life cycle parasites), but these carnivores may still be subject of parasite and predation risk from other carnivores. Herbivores may be at predation and parasitism risk at both carnivore and herbivore carcass sites, with the latter representing comparatively higher risk of acquiring parasites. Interactions among carnivores are more frequent at large carcasses [31] and in the absence of vultures [32], which are highly efficient scavengers [33,34]. Arrow width is roughly proportional to the intensity of risk.

Among the different pathways associated with carnivore carrion, only the risk of acquiring parasites through intra-guild and, especially, intra-specific scavenging received some scientific interest within the reviewed articles (e.g., [35–37]; Figure 3A).

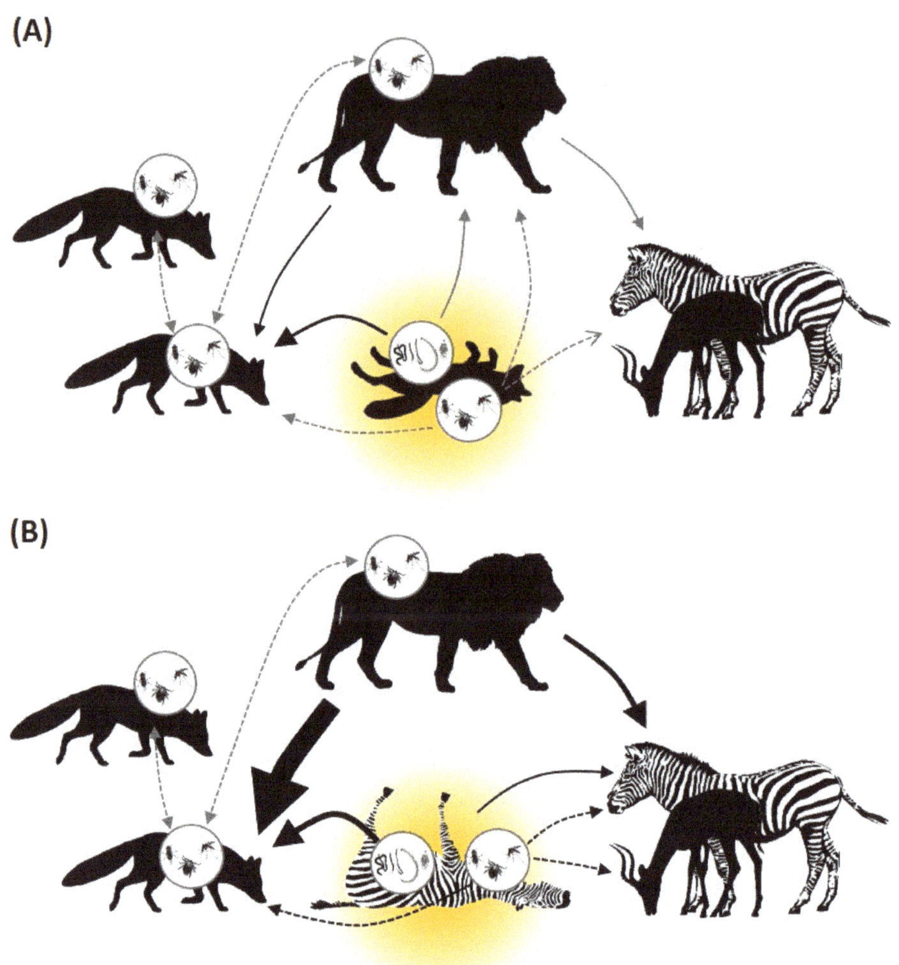

Figure 3. Research effort devoted to the main fear- and disgust-related interaction pathways around (**A**) carnivore and (**B**) herbivore carrion, according to our literature review. Those pathways for which we found at least one article are shown by black arrows (we use grey arrows otherwise). Arrow width is proportional to the number of articles.

Regarding herbivore carcasses (Figure 3B), most studies have focused on the predation risk affecting species that are present in the vicinities of carcass sites, both herbivores (e.g., [8,38–40]) and, mostly, carnivores (e.g., [9,41–47]). Within these articles, several studies explored how Plio/Pleistocene hominins were subject of predation risk while exploiting large carcasses (a process called confrontational/aggressive/active/power scavenging), according to paleontological evidences and behavioral, ecological, and energetics modelling (e.g., [13,48]), behavioral studies of modern human hunter-gatherer societies [49] or other primates [50], and other procedures [51]. Other studies have focused on several aspects of parasitism, especially the risk of acquiring parasites from meat consumption by carnivores (e.g., [52–54]).

3.3. Carrion Is Disgusting

Mammalian carnivores avoid feeding on other carnivore carcasses, especially of conspecifics, likely to reduce exposure to parasites [35–37]. These findings highlight that the risk of direct infection is higher among phylogenetically close organisms, which share more parasite species [1,29,30,37]. In turn, the carnivore carrion-avoidance behavior of carnivores enables a wide array of indirect ecological effects, both consumptive (e.g., scavenging of mammalian carnivore carcasses by a well-structured community of insects) and non-consumptive (e.g., hair collection by birds for nest construction), linked to carnivore carcasses [37,55]. This, together with the observation that grazers avoid foraging near herbivore carcasses to prevent infectious risk (e.g., [2,53,56]), indicates that dead animals may disgust both scavenging and non-scavenging species (Figure 3). However, effects of carrion on vegetation growth may lead to increased disease risk for herbivores that are attracted to carcass sites once carrion has been removed, in a sort of ecological trap that favors infection by highly resistant pathogens [11]. Regarding the mechanism by which scavengers may discriminate risky carcasses, some carnivores such as mammalian meso-carnivores seem to rely on carrion odor (e.g., to distinguish intra- from inter-specific carrion [37]), while others such as ants and beetles may detect the presence of certain pathogens by smelling or tasting their metabolites in the carcass [52].

In addition to meat-borne parasites, other pathogens present at carcass sites could be transmitted to any animal that approaches at a sufficient distance [11,53,57]. In addition, carcasses may indirectly favor parasite transmission among the scavengers that come into contact while scavenging, a circumstance that would be especially plausible in the absence of vultures [32], which are specialized, obligate scavengers that quickly remove carcasses [33,34,58]. However, these and other mechanisms of carcass-mediated infection risk (Figure 3) remain largely speculative and need further empirical support.

Besides carcass type, carrion-related infection risk—and the duration of the infective period—may be dependent on many other factors, such as parasite identity, carcass origin, the degree of starvation shown by the scavenger, and climatic conditions (e.g., [35,41,59]). Given the important ecological, evolutionary, and sanitary implications of the management of wild [60] and domestic animal carcasses [61], these issues require urgent scientific attention.

3.4. Carrion Is Scary

Herbivore carcass sites may be avoided by other herbivores because of high probability of predator–prey encounters (e.g., [8,38,40]; Figure 3B). Indeed, the probability of predation of ground-nesting birds increases near both predictable and unpredictable carrion resources [39,62]. The increased predation risk around carcasses is mainly explained by the fact that most predators behave also as facultative scavengers that are, to a greater or lesser extent, attracted to carrion [63,64]. This predation risk is likely higher at large carcasses because they are visited by more and larger predator species than smaller carcasses [7,31]. In addition to herbivores, small and medium-sized carnivores also avoid carcasses to prevent the risk of confronting dominant predators (e.g., [5,9,43,45,47,65]). In fact, the risk of being attacked by a larger predator, such as lion (*Panthera leo*) and spotted hyaena (*Crocuta crocuta*), may be so high that certain sympatric mammalian carnivores, such as cheetahs (*Acinonyx jubatus*) and wild dogs (*Lycaon pictus*), very rarely scavenge [66]. Cheetahs, even leave their own kills once satiated, no matter how much meat may be left [67].

However, how carnivores and herbivores behave at carnivore carcasses in relation to predation risk is virtually unknown (but see [41,65]; Figure 3A). In these cases, carcasses of predators may also be scary by themselves, as other scavengers that are within the prey base of the dead predators might even avoid the risk of inspecting such carcasses.

The risk that a given animal, either scavenger or not, is willing to accept at a carcass site may depend on several factors, such as sociality. For instance, spotted hyaenas are able to even displace lions from a carcass provided that the former outnumber the latter by a factor of 4 and no male lions are present at the carcass [68]. Coyotes (*Canis latrans*) may also displace wolves (*C. lupus*) from their kills when the former are in numerical advantage [9]. In addition, both hyaenas and (alpha) coyotes may trade off greater risk for high-quality carrion, as dominant carnivores may also facilitate carrion supply and detection [9,68]. Finally, hungry, sick, unexperienced, or senescent scavengers are probably more prone to face risky situations (e.g., [44,65,69]), though this needs further empirical confirmation.

3.5. Conclusions and Directions for Further Research

Overall, this review highlights the manifold inter- and intra-specific interactions linked to carrion via predation and parasitism risks, which may affect not only carrion consumers, but also non-consumers. Animal carcasses are an outstanding paradigm of how predation and parasitism pressures can concur in space and time, which is a major gap in predator- and parasite-avoidance scientific knowledge [1]. Carcasses may represent hotspots of infection and predation risk to both carnivores and herbivores, although the risk is highly dependent on a number of factors, such as carcass identity (Figure 2) and the size of the animals visiting carcass sites. For instance, large herbivores and top predators will be more reactive to parasite than to predator risk [1]. Thus, the multiple predator and parasite risk pathways that may arise around carnivore (Figure 2A) and herbivore carrion (Figure 2B), which are far more diverse than previously recognized, may differ qualitatively and quantitatively.

However, fully understanding of animal behavior around carrion resources requires exploring simultaneously different sources of risk [1], as avoiding one risk may increase [70] or decrease [71] another. Furthermore, Buck et al. [1] argue that "because predators are generally more mobile than parasites, the predator-induced landscape of fear might be more dynamic than the parasite-induced landscape of disgust". We highlight that, given the generally unpredictable and ephemeral nature of carrion [63,64,72], the very different life cycles of different parasite species [73], and the seasonality associated with their infective stages [74], carcass-induced landscape of disgust may be also highly dynamic.

The conceptual framework of the landscape of peril might also benefit from empirical evidence of aquatic ecosystems addressing key issues that differentiate them from terrestrial ones. For example, parasites in aquatic environments are more mobile than in terrestrial ecosystems, due to both active (i.e., locomotion and motility) and passive (e.g., currents and tides) transport through water [18]. Other relevant differences between terrestrial and aquatic systems relate to the cues to detect predators and parasites, which may differ qualitatively and quantitatively between air and water. In general, while visual, auditory and mechanosensory cues play a more prominent role in terrestrial environments, chemical cues are substantially more used by aquatic animals [18]. Comparative studies on how animals perceive and avoid predation and parasitism risk at carcass sites in terrestrial vs. aquatic environments are virtually absent, which opens exciting avenues for further research.

The conceptual model we present here (Figure 2) allows formulating predictions about how the landscape of fear and disgust around carcasses might influence animals' individual behavior and ecological processes, from population to ecosystem functioning. This could be especially useful in the current global change scenario, which includes high rates of species extinctions, invasions, and re-colonizations of both predators and parasites (e.g., [75]), as well as a growing evidence of the effect of human footprint on scavenger guilds [7]. In addition, our literature review has clearly shown that the research effort so far on predator and parasite risks associated with carrion has been highly unevenly distributed through the different interaction pathways, with most studies dealing with predation risk at vertebrate herbivore carcasses in terrestrial ecosystems (Figure 3). Thus, there is ample room and motivation for future investigation. Furthermore, most research has focused on behavioral responses (particularly, avoidance) of different species in relation to fear and disgust. However, to which extent are these behaviors innate or learned is an open question [36]. Moreover, other individual responses (e.g., physiological), as well as the effects at the population, community, and ecosystem levels remain largely unexplored and require further empirical evidence. While physiological responses have not been explored so far in a carrion context (to our knowledge), it is reasonable to think that the risks associated with carrion may exert different physiological costs (e.g., transitory and chronic stress) on animals visiting carcass sites, such as prey and subordinate predators. Future research might benefit from the application of novel (including experimental) methods in scavenging ecology and its interaction with different disciplines, as well as from the spatiotemporal quantification of carrion biomass [28,76] and the long-term monitoring of carcasses and scavenger guilds in different ecosystems [7]. Finally, besides freshwater and marine studies, terrestrial studies from tropical biomes would be especially welcome, as most research (for the historical period context) so far has focused on temperate, Mediterranean, and boreal systems.

In conclusion, future research should study the trade-offs and synergistic effects of both predator and parasite risk associated with carcasses of different nature and size in contrasting ecosystems and seasons, as well as the relative importance of these and other selective pressures. These ecological processes may have important consequences for animals facing predator and parasite risks, with individual costs ranging from diminished feeding rate to death, which may lead to wide ecological, evolutionary, epidemiological, forensic, and conservation implications. Acknowledging the multiple trophic (e.g., [64]) and non-trophic (e.g., [77]) ways in which dead animals directly and indirectly interact with living animals, including both herbivores and carnivores, will notably improve our understanding of the overlapping pressures that shape the landscape of fear and disgust.

Author Contributions: M.M. and J.A.S.-Z. conceived the study, conducted the literature review, wrote the manuscript and agreed to its submission. All authors have read and agreed to the published version of the manuscript.

Funding: M.M. was supported by a research contract Ramón y Cajal from the MINECO (RYC-2015-19231). This study was partly funded by the Spanish Ministry of Economy, Industry and Competitiveness and EU ERDF funds through the projects CGL2015-66966-C2-1-2-R and CGL2017-89905-R.

Conflicts of Interest: The authors declare no conflict of interest.

Appendix A References A1. Articles Reviewed. Articles Identified in the Google Scholar Search Only Are Marked with an Asterisk

1. *Allen, M.L.; Elbroch, L.M.; Wilmers, C.C.; Witmer, H.U. The comparative effects of large carnivores on the acquisition of carrion by scavengers. *Am. Nat.* **2015**, *185*, 822–833.
2. Allen, M.L.; Wilmers, C.C.; Elbroch, L.M.; Golla, J.M.; Witmer, H.U. The importance of motivation, weapons, and foul odors in driving encounter competition in carnivores. *Ecology* **2016**, *97*, 1905–1912.
3. *Archer, M.S.; Elgar, M.A. Effects of decomposition on carcass attendance in a guild of carrion-breeding flies. *Med. Vet. Entomol.* **2003**, *17*, 263–271.
4. *Archer, M.S.; Elgar, M.A. Female breeding-site preferences and larval feeding strategies of carrion-breeding Calliphoridae and Sarcophagidae (Diptera): A quantitative analysis. *Aust. J. Zool.* **2003**, *51*, 165–174.
5. Atwood, T.C.; Gese, E.M. Coyotes and recolonizing wolves: Social rank mediates risk-conditional behaviour at ungulate carcasses. *Anim. Behav.* **2008**, *75*, 753–762.
6. *Blanco, G.; Cardells, J.; Garijo-Toledo, M.M. Supplementary feeding and endoparasites in threatened avian scavengers: Coprologic evidence from red kites in their wintering stronghold. *Environ. Res.* **2017**, *155*, 22–30.
7. Blumenschine, R.J. Hominid carnivory and foraging strategies, and the socio-economic function of early archaeological sites. *Philos. Trans. R. Soc. Lond. B* **1991**, *334*, 211–221.
8. *Blumenschine, R.J.; Cavallo, J.A.; Capaldo, S.D. Competition for carcasses and early hominid behavioral ecology: A case study and conceptual framework. *J. Hum. Evol.* **1994**, *27*, 197–213.

9. Blumenschine, R.J.; Stanistreet, I.G.; Njau, J.K.; Bamford, M.K.; Masao, F.T.; Albert, R.M.; Stollhofen, H.; Andrews, P. Prassack, K.A.; McHenry, L.J.; et al. Environments and hominin activities across the FLK Peninsula during *Zinjanthropus* times (1.84 Ma), Olduvai Gorge, Tanzania. *J. Hum. Evol.* **2012**, *63*, 364–383.

10. *Cortés-Avizanda, A.; Selva, N.; Carrete, M.; Donázar, J.A. Effects of carrion resources on herbivore spatial distribution are mediated by facultative scavengers. *Basic Appl. Ecol.* **2009**, *10*, 265–272.

11. Cortés-Avizanda, A.; Carrete, M.; Serrano, D.; Donázar, J.A. Carcasses increase the probability of predation of ground-nesting birds: A caveat regarding the conservation value of vulture restaurants. *Anim. Conserv.* **2009**, *12*, 85–88.

12. Cunningham, C.X.; Johnson, C.N.; Barmuta, L.A.; Hollings, T.; Woehler, E.J.; Jones, M.E. Top carnivore decline has cascading effects on scavengers and carrion persistence. *Proc. R. Soc. B* **2018**, *285*, 20181582.

13. Daleo, P.; Alberti, J.; Avaca, M.S.; Narvarte, M.; Martinetto, M.; Iribarne, O. Avoidance of feeding opportunities by the whelk *Buccinanops globulosum* in the presence of damaged conspecifics. *Mar. Biol.* **2012**, *159*, 2359–2365.

14. *DeVault, T.L.; Rhodes, O.E., Jr. Identification of vertebrate scavengers of small mammal carcasses in a forested landscape. *Acta Theriol.* **2002**, *47*, 185–192.

15. Fialho, V.S.; Rodrigues, V.B.; Elliot, S.L. Nesting strategies and disease risk in necrophagous beetles. *Ecol. Evol.* **2018**, *8*, 3296–3310.

16. Fodrie, F.J.; Brodeur, M.C.; Toscano, B.J.; Powers, S.P. Friend or foe: Conflicting demands and conditional risk taking by opportunistic scavengers. *J. Exp. Mar. Biol. Ecol.* **2012**, *422–423*, 114–121.

17. *Foltan, P.; Puza, V. To complete their life cycle, pathogenic nematode-bacteria complexes deter scavengers from feeding on their host cadaver. *Behav. Process.* **2009**, *80*, 76–79.

18. Frank, S.C.; Blaalid, R.; Mayer, M.; Zedrosser, A.; Steyaert, S.M.J.G. Fear the reaper: Ungulate carcasses may generate an ephemeral landscape of fear for rodents. *R. Soc. Open Sci.* **2020**, *7*, 191644.

19. García-García, F.J.; Reyes-Martínez, M.J.; Ruiz-Delgado, M.C.; Sánchez-Moyano, J.E.; Castro-Casas, M.; Pérez-Hurtado, A. Does the gathering of shellfish affect the behavior of gastropod scavengers on sandy beaches? A field experiment. *J. Exp. Mar. Biol. Ecol.* **2015**, *467*, 1–6.

20. Halley, D.J. Interspecific dominance and risk-taking in three species of corvid scavenger. *J. Yamashina Inst. Ornithol.* **2001**, *33*, 44–50.

21. *Hunter, J.S.; Durant, S.M.; Caro, T.M. To flee or not to flee: Predator avoidance by cheetahs at kills. *Behav. Ecol. Sociob.* **2007**, *61*, 1033–1042.

22. *Jennelle, C.S.; Samuel, M.D.; Nolden, C.A.; Berkley, E.A. Deer carcass decomposition and potential scavenger exposure to chronic wasting disease. *J. Wildl. Manage.* **2009**, *73*, 655–662.

23. Jones, M.E. The function of vigilance in sympatric marsupial carnivores: The eastern quoll and the Tasmanian devil. *Anim. Behav.* **1998**, *56*, 1279–1284.

24. *Lev-Yadun, S.; Gutman, M. Carrion odor and cattle grazing. *Comm. Integr. Biol.* **2013**, *6*, e26111.

25. *Lupo, K.D. Experimentally derived extraction rates for marrow: Implications for body part exploitation strategies of Plio-Pleistocene hominid scavengers. *J. Archaeol. Sci.* **1998**, *25*, 657–675.

26. *Mattisson, J.; Rauset, G.R.; Odden, J.; Andrén, H.; Linnell, J.D.C.; Persson, J. Predation or scavenging? Prey body condition influences decision-making in a facultative predator, the wolverine. *Ecosphere* **2016**, *7*, e01407.

27. McKillup, S.C.; McKillup, R.V. The decision to feed by a scavenger in relation to the risks of predation and starvation. *Oecologia* **1994**, *97*, 41–48.

28. *Moleón, M.; Sánchez-Zapata, J.A.; Sebastián-González, E.; Owen-Smith, N. Carcass size shapes the structure and functioning of an African scavenging assemblage. *Oikos* **2015**, *124*, 1391–1403.

29. Moleón, M.; Martínez-Carrasco, C.; Muellerklein, O.C.; Getz, W.M.; Muñoz-Lozano, C.; Sánchez-Zapata, J.A. Carnivore carcasses are avoided by carnivores. *J. Anim. Ecol.* **2017**, *86*, 1179–1191.

30. Monahan, C.M. Tha Hadza carcass transport debate revisited and its archaeological implications. *J. Archaeol. Sci.* **1998**, *25*, 405–424.

31. Morton, B.; Chan, K. Hunger rapidly overrides the risk of predation in the subtidal scavenger *Nassarius siquijorensis* (Gastropoda: Nassariidae): An energy budget and a comparison with the intertidal *Nassarius festivus* in Hong Kong. *J. Exp. Mar. Biol. Ecol.* **1999**, *240*, 213–228.

32. *Muñoz-Lozano, C.; Martín-Vega, D.; Martínez-Carrasco, C.; Sánchez-Zapata, J.A.; Morales-Reyes, Z.; Gonzálvez, M.; Moleón, M. Avoidance of carnivore carcasses by vertebrate scavengers enables colonization by a diverse community of carrion insects. *PLoS ONE* **2019**, *14*, e0221890.

33. *Navarro, F.K.S.P.; Rezende, R. de S.; Gonçalves, J.F., Jr. Experimental assessment of temperature increase and presence of predator carcass changing the response of invertebrate shredders. *Biota Neotrop.* **2013**, *13*.

34. Oliver, J.S. Estimates of hominid and carnivore involvement in the FLK *Zinjanthropus* fossil assemblage: Some socioecological implications. *J. Hum. Evol.* **1994**, *27*, 267–294.
35. *Olson, Z.H.; Beasley, J.C.; Rhodes, O.E., Jr. Carcass type affects local scavenger guilds more than habitat connectivity. *PLoS ONE* **2016**, *11*, e0147798.
36. O'Malley, C.; Elbroch, L.M.; Lendrum, P.E.; Quigley, H. Motion-triggered video cameras reveal spatial and temporal patterns of red fox foraging on carrion provided by mountain lions. *PeerJ* **2018**, *6*, e5324.
37. *Pereira, H.; Detrain, C. Pathogen avoidance and prey discrimination in ants. *R. Soc. Open Sci.* **2020**, *7*, 191705.
38. *Ragir, S.; Rosenberg, M.; Tierno, P. Gut morphology and the avoidance of carrion among chimpanzees, baboons, and early hominids. *J. Anthropol. Res.* **2000**, *56*, 477–512.
39. *Rees, J.D.; Webb, J.K.; Crowther, M.S.; Letnic, M. Carrion subsidies provided by fishermen increase predation of beach-nesting bird nests by facultative scavengers. *Anim. Conserv.* **2015**, *18*, 44–49.
40. Rees, J.D., Crowther, M.S., Kingsford, R.T., Letnic, M. Direct and indirect effects of carrion subsidies in an arid rangeland: Carrion has positive effects on facultative scavengers and negative effects on a small songbird. *J. Arid Environ.* **2020**, *179*, 104174.
41. Rose, L.; Marshall, F. Meat eating, hominid sociality, and home bases revisited. *Curr. Anthropol.* **1996**, *37*, 307–338.
42. *Schlacher, T.A.; Strydom, S.; Connolly, R.M. Multiple scavengers respond rapidly to pulsed carrion resources at the land-ocean interface. *Acta Oecol.* **2013**, *48*, 7–12.
43. Schlacher, T.A.; Weston, M.A.; Lynn, D.; Schoeman, D.S.; Huijbers, C.M.; Olds, A.D.; Masters, S.; Connolly, R.M. Conservation gone to the dogs: When canids rule the beach in small coastal reserves. *Biodivers. Conserv.* **2015**, *24*, 493–509.
44. *Selva, N.; Jędrzejewska, B.; Jędrzejewski, W.; Wajrak, A. Factors affecting carcass use by a guild of scavengers in European temperate woodland. *Can. J. Zool.* **2005**, *83*, 1590–1601.
45. *Siva-Jothy, J.A.; Monteith, K.M.; Vale, P.F. Navigating infection risk during oviposition and cannibalistic foraging in a holometabolous insect. *Behav. Ecol.* **2018**, *29*, 1426–1435.
46. Steinbeiser, C.M.; Wawrzynowski, C.A.; Ramos, X.; Olson, Z.H. Scavenging and the ecology of fear: Do animal carcasses create islands of risk on the landscape? *Can. J. Zool.* **2017**, *96*, 229–236.
47. *Tranter, C.; LeFevre, L.; Evison, S.E.F.; Hughes, W.O.H. Threat detection: Contextual recognition and response to parasites by ants. *Behav. Ecol.* **2015**, *26*, 396–405.
48. Turner, W.C.; Kausrud, K.L.; Krishnappa, Y.S.; Cromsigt, J.P.G.M.; Ganz, H.H.; Mapaure, I.; Cloete, C.C.; Havarua, Z.; Küsters, M.; Getz, W.M.; et al. Fatal attraction: Vegetation responses to nutrient inputs attract herbivores to infectious anthrax carcass sites. *Proc. R. Soc. B* **2014**, *281*, 20141785.
49. *van Dijk, J.; Andersen, T.; May, R.; Andersen, R.; Andersen, R.; Landa, A. Foraging strategies of wolverines within a predator guild. *Can. J. Zool.* **2008**, *86*, 966–975.
50. *Wikenros, C.; Sand, H.; Ahlqvist, P.; Liberg, O. Biomass flow and scavengers use of carcasses after re-colonization of an apex predator. *PLoS ONE* **2013**, *8*, e77373.
51. Wikenros, C.; Ståhlberg, S.; Sand, H. Feeding under high risk of intraguild predation: Vigilance patterns of two medium-sized generalist predators. *J. Mammal.* **2014**, *95*, 862–870.
52. Willems, E.P.; van Schaik, C.P. The social organization of *Homo ergaster*: Inferences from anti-predator responses in extant primates. *J. Hum. Evol.* **2017**, *109*, 11–21.

References

1. Buck, J.C.; Weinstein, S.B.; Young, H.S. Ecological and evolutionary consequences of parasite avoidance. *Trends Ecol. Evol.* **2018**, *33*, 619–632. [CrossRef] [PubMed]
2. Weinstein, S.B.; Buck, J.C.; Young, H.S. A landscape of disgust. *Science* **2018**, *359*, 1213–1214. [CrossRef]
3. Clinchy, M.; Sheriff, M.J.; Zanette, L.Y. Predator-induced stress and the ecology of fear. *Funct. Ecol.* **2013**, *27*, 56–65. [CrossRef]
4. Sarabian, C.; Curtis, V.; McMullan, R. Evolution of pathogen and parasite avoidance behaviours. *Phil. Trans. R. Soc. B* **2018**, *373*, 20170256. [CrossRef] [PubMed]
5. Cunningham, C.X.; Johnson, C.N.; Barmuta, L.A.; Hollings, T.; Woehler, E.J.; Jones, M.E. Top carnivore decline has cascading effects on scavengers and carrion persistence. *Proc. R. Soc. B* **2018**, *285*, 20181582. [CrossRef]
6. Doherty, J.-F.; Ruehle, B. An integrated landscape of fear and disgust: The evolution of avoidance behaviors amidst a myriad of natural enemies. *Front. Ecol. Evol.* **2020**, *8*, 564343. [CrossRef]
7. Sebastián-González, E.; Barbosa, J.M.; Pérez-García, J.M.; Morales-Reyes, Z.; Botella, F.; Olea, P.; Mateo-Tomás, P.; Moleón, M.; Hiraldo, F.; Arrondo, E.; et al. Scavenging in the Anthropocene: Human impact drives vertebrate scavenger species richness at a global scale. *Glob. Chang. Biol.* **2019**, *25*, 3005–3017. [CrossRef] [PubMed]
8. Frank, S.C.; Blaalid, R.; Mayer, M.; Zedrosser, A.; Steyaert, S.M.J.G. Fear the reaper: Ungulate carcasses may generate an ephemeral landscape of fear for rodents. *R. Soc. Open Sci.* **2020**, *7*, 191644. [CrossRef]

9. Atwood, T.C.; Gese, E.M. Coyotes and recolonizing wolves: Social rank mediates risk-conditional behaviour at ungulate carcasses. *Anim. Behav.* **2008**, *75*, 753–762. [CrossRef]

10. Blanco, G.; Cardells, J.; Garijo-Toledo, M.M. Supplementary feeding and endoparasites in threatened avian scavengers: Coprologic evidence from red kites in their wintering stronghold. *Environ. Res.* **2017**, *155*, 22–30. [CrossRef] [PubMed]

11. Turner, W.C.; Kausrud, K.L.; Krishnappa, Y.S.; Cromsigt, J.P.G.M.; Ganz, H.H.; Mapaure, I.; Cloete, C.C.; Havarua, Z.; Küsters, M.; Getz, W.M.; et al. Fatal attraction: Vegetation responses to nutrient inputs attract herbivores to infectious anthrax carcass sites. *Proc. R. Soc. B* **2014**, *281*, 20141785. [CrossRef] [PubMed]

12. Ragir, S.; Rosenberg, M.; Tierno, P. Gut morphology and the avoidance of carrion among chimpanzees, baboons, and early hominids. *J. Anthropol. Res.* **2000**, *56*, 477–512. [CrossRef]

13. Blumenschine, R.J.; Stanistreet, I.G.; Njau, J.K.; Bamford, M.K.; Masao, F.T.; Albert, R.M.; Stollhofen, H.; Andrews, P.; Prassack, K.A.; McHenry, L.J.; et al. Environments and hominin activities across the FLK Peninsula during *Zinjanthropus* times (1.84 Ma), Olduvai Gorge, Tanzania. *J. Hum. Evol.* **2012**, *63*, 364–383. [CrossRef] [PubMed]

14. Moleón, M.; Sánchez-Zapata, J.A.; Margalida, A.; Carrete, M.; Owen-Smith, N.; Donázar, J.A. Humans and scavengers: The evolution of interactions and ecosystem services. *BioScience* **2014**, *64*, 394–403. [CrossRef]

15. Lev-Yadun, S.; Ne'eman, G.; Shanas, U. A sheep in wolf's clothing: Do carrion and dung odours of flowers not only attract pollinators but also deter herbivores? *BioEssays* **2009**, *31*, 84–88. [CrossRef]

16. Evans, T.S.; Shi, Z.; Boots, M.; Liu, W.; Olival, K.J.; Xiao, X.; Vandewoude, S.; Brown, H.; Chen, J.L.; Civitello, D.J.; et al. Synergistic China-US ecological research is essential for global emerging infectious disease preparedness. *EcoHealth* **2020**, *17*, 160–173. [CrossRef]

17. Getz, W.M. Biomass transformation webs provide a unified approach to consumer-resource modelling. *Ecol. Lett.* **2011**, *14*, 113–124. [CrossRef]

18. Behringer, D.C.; Karvonen, A.; Bojko, J. Parasite avoidance behaviours in aquatic environments. *Philos. Trans. R. Soc. B* **2018**, *373*, 20170202. [CrossRef] [PubMed]

19. Haddaway, N.R.; Woodcock, P.; Macura, B.; Collins, A. Making literature reviews more reliable through application of lessons from systematic reviews. *Conserv. Biol.* **2015**, *29*, 1596–1605. [CrossRef]

20. Lozano, J.; Olszańska, A.; Morales-Reyes, Z.; Castro, A.A.; Malo, A.F.; Moleón, M.; Sánchez-Zapata, J.A.; Cortés-Avizanda, A.; von Wehrden, H.; Dorresteijn, I.; et al. Human-carnivore relations: A systematic review. *Biol. Conserv.* **2019**, *237*, 480–492. [CrossRef]

21. Lima, S.L.; Dill, L.M. Behavioral decisions made under the risk of predation: A review and prospectus. *Can. J. Zool.* **1990**, *68*, 619–640. [CrossRef]

22. Rohr, J.R.; Swan, A.; Raffel, T.R.; Hudson, P.J. Parasites, info-disruption, and the ecology of fear. *Oecologia* **2009**, *159*, 447–454. [CrossRef] [PubMed]

23. Blumstein, D.T.; Rangchi, T.N.; Briggs, T.; Souza de Andrade, F.; Natterson-Horowitz, B. A systematic review of carrion eater's adaptations to avoid sickness. *J. Wildl. Dis.* **2017**, *53*, 577–581. [CrossRef] [PubMed]

24. Baruzzi, C.; Mason, D.; Barton, B.; Lashley, M. Effects of increasing carrion biomass on food webs. *Food Webs* **2018**, *16*, e00096. [CrossRef]

25. Buck, J.C. Indirect effects explain the role of parasites in ecosystems. *Trends Parasitol.* **2019**, *35*, 835–847. [CrossRef]

26. Case, T.I.; Stevenson, R.J.; Byrne, R.W.; Hobaiter, C. The animal origins of disgust: Reports of basic disgust in nonhuman great apes. *Evol. Behav. Sci.* **2020**, *14*, 231–260. [CrossRef]

27. Prugh, L.R.; Sivy, K.J. Enemies with benefits: Integrating positive and negative interactions among terrestrial carnivores. *Ecol. Lett.* **2020**, *23*, 902–918. [CrossRef] [PubMed]

28. Moleón, M.; Selva, N.; Sánchez-Zapata, J.A. The components and spatiotemporal dimensions of carrion biomass quantification. *Trends Ecol. Evol.* **2020**, *35*, 91–92. [CrossRef]

29. Huang, S.; Bininda-Emonds, O.R.; Stephens, O.R.; Gittleman, J.L.; Altizer, S. Phylogenetically related and ecologically similar carnivores harbor similar parasite assemblages. *J. Anim. Ecol.* **2014**, *83*, 671–680. [CrossRef] [PubMed]

30. Stephens, P.R.; Altizer, S.; Smith, K.F.; Aguirre, A.A.; Brown, J.H.; Budischak, S.A.; Byers, J.E.; Dallas, T.A.; Davies, T.J.; Drake, J.M.; et al. The macroecology of infectious diseases: A new perspective on global-scale drivers of pathogen distributions and impacts. *Ecol. Lett.* **2016**, *19*, 1159–1171. [CrossRef] [PubMed]

31. Moleón, M.; Sánchez-Zapata, J.A.; Sebastián-González, E.; Owen-Smith, N. Carcass size shapes the structure and functioning of an African scavenging assemblage. *Oikos* **2015**, *124*, 1391–1403. [CrossRef]

32. Ogada, D.L.; Torchin, M.E.; Kinnaird, M.F.; Ezenwa, V.O. Effects of vulture declines on facultative scavengers and potential implications for mammalian disease transmission. *Conserv. Biol.* **2012**, *26*, 453–460. [CrossRef]

33. Morales-Reyes, Z.; Sánchez-Zapata, J.A.; Sebastián-González, E.; Botella, F.; Carrete, M.; Moleón, M. Scavenging efficiency and red fox abundance in Mediterranean mountains with and without vultures. *Acta Oecol.* **2017**, *79*, 81–88. [CrossRef]

34. Hill, J.E.; DeVault, T.L.; Beasley, J.C.; Rhodes, O.E., Jr.; Belant, J.L. Effects of vulture exclusion on carrion consumption by facultative scavengers. *Ecol. Evol.* **2018**, *8*, 2518–2526. [CrossRef] [PubMed]

35. Selva, N.; Jędrzejewska, B.; Jędrzejewski, W.; Wajrak, A. Factors affecting carcass use by a guild of scavengers in European temperate woodland. *Can. J. Zool.* **2005**, *83*, 1590–1601. [CrossRef]

36. Olson, Z.H.; Beasley, J.C.; Rhodes, O.E., Jr. Carcass type affects local scavenger guilds more than habitat connectivity. *PLoS ONE* **2016**, *11*, e0147798. [CrossRef] [PubMed]

37. Moleón, M.; Martínez-Carrasco, C.; Muellerklein, O.C.; Getz, W.M.; Muñoz-Lozano, C.; Sánchez-Zapata, J.A. Carnivore carcasses are avoided by carnivores. *J. Anim. Ecol.* **2017**, *86*, 1179–1191. [CrossRef] [PubMed]

38. Cortés-Avizanda, A.; Selva, N.; Carrete, M.; Donázar, J.A. Effects of carrion resources on herbivore spatial distribution are mediated by facultative scavengers. *Basic Appl. Ecol.* **2009**, *10*, 265–272. [CrossRef]

39. Cortés-Avizanda, A.; Carrete, M.; Serrano, D.; Donázar, J.A. Carcasses increase the probability of predation of ground-nesting birds: A caveat regarding the conservation value of vulture restaurants. *Anim. Conserv.* **2009**, *12*, 85–88. [CrossRef]

40. Steinbeiser, C.M.; Wawrzynowski, C.A.; Ramos, X.; Olson, Z.H. Scavenging and the ecology of fear: Do animal carcasses create islands of risk on the landscape? *Can. J. Zool.* **2017**, *96*, 229–236. [CrossRef]

41. McKillup, S.C.; McKillup, R.V. The decision to feed by a scavenger in relation to the risks of predation and starvation. *Oecologia* **1994**, *97*, 41–48. [CrossRef]

42. Archer, M.S.; Elgar, M.A. Effects of decomposition on carcass attendance in a guild of carrion-breeding flies. *Med. Vet. Entomol.* **2003**, *17*, 263–271. [CrossRef] [PubMed]

43. Van Dijk, J.; Andersen, T.; May, R.; Andersen, R.; Andersen, R.; Landa, A. Foraging strategies of wolverines within a predator guild. *Can. J. Zool.* **2008**, *86*, 966–975. [CrossRef]

44. Fodrie, F.J.; Brodeur, M.C.; Toscano, B.J.; Powers, S.P. Friend or foe: Conflicting demands and conditional risk taking by opportunistic scavengers. *J. Exp. Mar. Biol. Ecol.* **2012**, *422–423*, 114–121. [CrossRef]

45. Schlacher, T.A.; Strydom, S.; Connolly, R.M. Multiple scavengers respond rapidly to pulsed carrion resources at the land-ocean interface. *Acta Oecol.* **2013**, *48*, 7–12. [CrossRef]

46. Wikenros, C.; Ståhlberg, S.; Sand, H. Feeding under high risk of intraguild predation: Vigilance patterns of two medium-sized generalist predators. *J. Mammal.* **2014**, *95*, 862–870. [CrossRef]

47. Allen, M.L.; Elbroch, L.M.; Wilmers, C.C.; Witmer, H.U. The comparative effects of large carnivores on the acquisition of carrion by scavengers. *Am. Nat.* **2015**, *185*, 822–833. [CrossRef]

48. Blumenschine, R.J. Hominid carnivory and foraging strategies, and the socio-economic function of early archaeological sites. *Philos. Trans. R. Soc. Lond. B* **1991**, *334*, 211–221.

49. Monahan, C.M. Tha Hadza carcass transport debate revisited and its archaeological implications. *J. Archaeol. Sci.* **1998**, *25*, 405–424. [CrossRef]

50. Willems, E.P.; van Schaik, C.P. The social organization of *Homo ergaster*: Inferences from anti-predator responses in extant primates. *J. Hum. Evol.* **2017**, *109*, 11–21. [CrossRef] [PubMed]

51. Lupo, K.D. Experimentally derived extraction rates for marrow: Implications for body part exploitation strategies of Plio-Pleistocene hominid scavengers. *J. Archaeol. Sci.* **1998**, *25*, 657–675. [CrossRef]

52. Foltan, P.; Puza, V. To complete their life cycle, pathogenic nematode-bacteria complexes deter scavengers from feeding on their host cadaver. *Behav. Process.* **2009**, *80*, 76–79. [CrossRef]

53. Jennelle, C.S.; Samuel, M.D.; Nolden, C.A.; Berkley, E.A. Deer carcass decomposition and potential scavenger exposure to chronic wasting disease. *J. Wildl. Manag.* **2009**, *73*, 655–662. [CrossRef]

54. Fialho, V.S.; Rodrigues, V.B.; Elliot, S.L. Nesting strategies and disease risk in necrophagous beetles. *Ecol. Evol.* **2018**, *8*, 3296–3310. [CrossRef]

55. Muñoz-Lozano, C.; Martín-Vega, D.; Martínez-Carrasco, C.; Sánchez-Zapata, J.A.; Morales-Reyes, Z.; Gonzálvez, M.; Moleón, M. Avoidance of carnivore carcasses by vertebrate scavengers enables colonization by a diverse community of carrion insects. *PLoS ONE* **2019**, *14*, e0221890. [CrossRef] [PubMed]

56. Lev-Yadun, S.; Gutman, M. Carrion odor and cattle grazing. *Comm. Integr. Biol.* **2013**, *6*, e26111. [CrossRef] [PubMed]

57. Lepczyk, C.A.; Lohr, C.A.; Duffy, D.C. A review of cat behavior in relation to disease risk and management options. *Appl. Anim. Behav. Sci.* **2015**, *173*, 29–39. [CrossRef]

58. Mateo-Tomás, P.; Olea, P.P.; Moleón, M.; Selva, N.; Sánchez-Zapata, J.A. Both rare and common species support ecosystem services in scavenging communities. *Glob. Ecol. Biogeogr.* **2017**, *26*, 1459–1470. [CrossRef]

59. Rossi, L.; Interisano, M.; Deksne, G.; Pozio, E. The subnivium, a haven for *Trichinella* larvae in host carcasses. *Int. J. Parasitol.* **2019**, *8*, 229–233. [CrossRef]

60. Margalida, A.; Moleón, M. Toward carrion-free ecosystems? *Front. Ecol. Environ.* **2016**, *14*, 183–184. [CrossRef]

61. Morales-Reyes, Z.; Pérez-García, J.M.; Moleón, M.; Botella, F.; Carrete, M.; Donázar, J.A.; Cortés-Avizanda, A.; Arrondo, E.; Moreno-Opo, R.; Jiménez, J.; et al. Evaluation of the network of protection areas for the feeding of scavengers in Spain: From biodiversity conservation to greenhouse gas emission savings. *J. Appl. Ecol.* **2017**, *54*, 1120–1129. [CrossRef]

62. Rees, J.D.; Webb, J.K.; Crowther, M.S.; Letnic, M. Carrion subsidies provided by fishermen increase predation of beach-nesting bird nests by facultative scavengers. *Anim. Conserv.* **2015**, *18*, 44–49. [CrossRef]

63. DeVault, T.L.; Rhodes, O.E., Jr.; Shivik, J.A. Scavenging by vertebrates: Behavioral, ecological, and evolutionary perspectives on an important energy transfer pathway in terrestrial ecosystems. *Oikos* **2003**, *102*, 225–234. [CrossRef]

64. Moleón, M.; Sánchez-Zapata, J.A.; Selva, N.; Donázar, J.A.; Owen-Smith, N. Inter-specific interactions linking predation and scavenging in terrestrial vertebrate assemblages. *Biol. Rev.* **2014**, *89*, 1042–1054. [CrossRef]

65. Morton, B.; Chan, K. Hunger rapidly overrides the risk of predation in the subtidal scavenger *Nassarius siquijorensis* (Gastropoda Nassariidae): An energy budget and a comparison with the intertidal *Nassarius festivus* in Hong Kong. *J. Exp. Mar. Biol. Ecol* **1999**, *240*, 213–228. [CrossRef]

66. Pereira, L.M.; Owen-Smith, N.; Moleón, M. Facultative predation and scavenging by mammalian carnivores: Seasonal, regional and intra-guild comparisons. *Mammal Rev.* **2014**, *44*, 44–45. [CrossRef]

67. Hunter, J.S.; Durant, S.M.; Caro, T.M. To flee or not to flee: Predator avoidance by cheetahs at kills. *Behav. Ecol. Sociobiol.* **2007**, *61*, 1033–1042. [CrossRef]

68. Amorós, M.; Gil-Sánchez, J.M.; López-Pastor, B.D.L.N.; Moleón, M. Hyaenas and lions: How the largest African carnivores interact at carcasses. *Oikos* **2020**, *129*, 1820–1832. [CrossRef]

69. Daleo, P.; Alberti, J.; Avaca, M.S.; Narvarte, M.; Martinetto, M.; Iribarne, O. Avoidance of feeding opportunities by the whelk *Buccinanops globulosum* in the presence of damaged conspecifics. *Mar. Biol.* **2012**, *159*, 2359–2365. [CrossRef]

70. Koprivnikar, J.; Penalva, L. Lesser of two evils? Foraging choices in response to threats of predation and parasitism. *PLoS ONE* **2015**, *10*, e0116569. [CrossRef] [PubMed]

71. Parsons, M.H.; Blumstein, D.T. Familiarity breeds contempt: Kangaroos persistently avoid areas with experimentally deployed dingo scents. *PLoS ONE* **2010**, *5*, e10403. [CrossRef] [PubMed]

72. Barton, P.S.; Cunningham, S.A.; Lindenmayer, D.B.; Manning, A.D. The role of carrion in maintaining biodiversity and ecological processes in terrestrial ecosystems. *Oecologia* **2013**, *171*, 761–772. [CrossRef] [PubMed]

73. Combes, C. *Parasitism: The Ecology and Evolution of Intimate Interactions*; The University of Chicago Press: Chicago, IL, USA, 2001.

74. Altizer, S.; Dobson, A.; Hosseini, P.; Hudson, P.; Pascual, M.; Rohani, P. Seasonality and the dynamics of infectious diseases. *Ecol. Lett.* **2006**, *9*, 467–484. [CrossRef]

75. Costard, S.; Wieland, B.; de Glanville, W.; Jori, F.; Rowlands, R.; Vosloo, W.; Roger, F.; Pfeiffer, D.U.; Dixon, L.K. African swine fever: How can global spread be prevented? *Philos. Trans. R. Soc. B* **2009**, *364*, 2683–2696. [CrossRef] [PubMed]

76. Barton, P.S.; Evans, M.J.; Foster, C.N.; Pechal, J.L.; Bump, J.K.; Quaggioto, M.-M.; Benbow, M.E. Towards quantifying carrion biomass in ecosystems. *Trends Ecol. Evol.* **2019**, *34*, 950–961. [CrossRef] [PubMed]

77. Moleón, M.; Sánchez-Zapata, J.A. Non-trophic functions of carcasses: From death to the nest. *Front. Ecol. Environ.* **2016**, *14*, 340–341. [CrossRef]

MDPI

St. Alban-Anlage 66

4052 Basel

Switzerland

Tel. +41 61 683 77 34

Fax +41 61 302 89 18

www.mdpi.com

Diversity Editorial Office

E-mail: diversity@mdpi.com

www.mdpi.com/journal/diversity

www.ingramcontent.com/pod-product-compliance
Lightning Source LLC
LaVergne TN
LVHW070616170726
843515LV00004B/99